化学英語の
スタイルガイド

㈳日本
松永

朝倉書

推薦の辞

　初めて英語論文を書こうとしても，なかなかできるものではない．化学論文の英語は日常の英語とは異なるうえ，雑誌ごとに書式も異なることが心理的な障壁になっていることが多い．私自身，日本化学会が出版している欧文誌 The Bulletin of the Chemical Society of Japan ならびに速報誌 Chemistry Letters については，『投稿規則・投稿の手引き』（日本化学会編，第 4 版，1989 年）を長らく参考にしてきたが，電子投稿の時代になって書式の大幅な変更もあり，もはや間に合わなくなってきた．2004 年 4 月から速報誌の編集委員長を務めるようになり，出版の過程をつぶさに知るにいたって問題意識をもっていたところ，速報誌編集顧問として論文の最終段階で英文ならびに書式の最終チェックをしていただいている松永義夫先生が，荒木啓介氏（情報言語研究所）の協力を得られて本書を出版されることになった．化学および広く化学にかかわる科学分野において，英語論文執筆時に必須の知識が，学生・院生ならびに研究者用にわかりやすく解説されている．とくに，論文で頻出する英語語句について，日本人研究者が英語論文の執筆で間違えやすいポイントを，アルファベット順位に，科学英文での実例，文法上・論文使用にあたっての注意，間違えやすい点などを併記して解説してある．一見して有用な参考書であることがわかる．しかも，日本化学会の 2 誌のみならずアメリカ化学会の雑誌に投稿する場合にはこうしたらよいと解説もしてある．日本化学会が自信をもって推奨する本である．ぜひとも化学研究に携わる多くの方に読んでいただき，論文投稿の際に活用していただきたいと願っている．

　2006 年 1 月

速報誌編集委員長　檜山爲次郎

まえがき

　本スタイルガイドは日本化学会の速報誌 Chemistry Letters に論文を投稿するに際して，注意すべき事項や英語の知識をまとめたものである．

　米式英語によって，文法上誤りがなく，化学用語が正しく使用された原稿を提出せよと言うだけでは，具体的にどうすればよいかが，著者に明らかであるとは言い難い．日本化学会の立場から見た原稿の完成には，欧文誌，速報誌両編集委員会の議を経た『投稿規則・投稿の手引』の参照が不可欠である．しかし，小冊子であり，長らく改定されていないこともあって，十分な情報は得られない．手近な論文誌を参照すると，米式か，英式かだけで，綴りのみならず，字体，コンマ，ハイフン，記号の使用にも，かなりの差異がある．さらに，論文誌ごとのスタイルの差異がこれにつけ加わる．日本化学会刊行の書籍，例えば『学術用語集』を参照して化学用語におけるハイフンの有無を知ろうとしても，必ずしも掲載されていない．有力な手段として，米国化学会発行の分厚い The ACS Style Guide を参考にすることが考えられる．しかし，日本化学会の『投稿規則・投稿の手引』や刊行物を無視して，全面的に The ACS Style Guide に頼ることはできない．記号「～」一つを取り上げても，『化学便覧基礎編』に記載されている国際的取り決めに従った用法は，The ACS Style Guide 記載の用法とはまったく異なる．したがって，日本化学会においても，原稿を作成するのに必要な事項を説明した解説書が存在することが望ましい．この要望にこたえて企画されたのが，速報誌の現状をまとめ，解説と例文を付した本スタイルガイドである．現在，速報誌のスタイルと欧文誌のスタイルには，『投稿規定・投稿の手引き』からも知れるように，多少の形式上の相違はあるが，それらは化学会学術情報部で整える範囲内にあるから，欧文誌に投稿する著者にも，本書は十分に役立つものである．

　The ACS Style Guide は数少ない章から成り立ち，区分けが大きくて，必要な箇所を見出すのが容易でなく，索引を見ると何ヵ所にも分散していることが，しばしばである．そこで，本書は慣用法辞典 Usage の形式を採用して，索引の項目程度に区分けを小さくするとともに，各項目に関連することは，多

少は重複しても一括して記載することによって，必要な事項に容易に到達できるよう心掛けた．日本人向けの本書には，The ACS Style Guide よりも数多くの英文に関する注意事項を取り上げ，文法にふれた詳しい解説がなされている．多数の例文は後述のように本書独自のものである．

本文は英語の見出しとして ABC 順に配列されているので，必要な項目を英語で直接参照することができるが，内容的な体系性は有していない．また，もとより英和辞典ではないので，多くの注意すべき表現を挙げているとはいえ，実際にどれが存在するかは見えにくい．そこで，日本語からのアプローチも可能にする目的に併せて，索引の主見出しと副見出し（説明句）を設け，本文中の数箇所で述べられている関連事項がまとまって探せるように工夫した．また必要に応じて同一の項目につき主見出しと副見出しを逆転させ二重に設定しているので，どちらからでも関連事項にたどり着ける．さらに索引全体にある程度の体系性を持たせ，説明箇所の構成や分量を見渡せるように努めた．

なお，本書は日本化学会学術情報部のご諒承をえて，化学会監修として出版されることとなった．特に檜山為次郎速報誌編集委員長，玉尾皓平欧文誌編集委員長には本書の出版にあたりご理解とご支援を頂いた．ここに厚くお礼申し上げる．また本書の企画立ち上げから出版まで，関係者間の調整を引き受けられた林和弘化学会学術情報部課長にも謝意を表する．

平素ともに仕事をしている荒木啓介，大橋守両顧問には，原稿について種々のご助言を頂いた．詳細な索引は荒木啓介顧問が使用者の便宜を考慮して，工夫し作成されたものである．両顧問のご協力に深く感謝する．また，刊行に当たって大変お世話になった朝倉書店編集部の方々に厚くお礼を申し上げる．

2005 年 12 月

速報誌製作顧問　松永義夫

使用の手引き

1 項目は英語とし，配列はアルファベット順による．
2 項目の多くには例文が付されている．
3 本書の利用を助けるため，主だった項目を紹介する．
   3.1 論文の構成にかかわる項目としては，次のものがある．

      caption　　　　　　　citation in text　　　conclusion
      references and notes　table　　　　　　　title

   3.2 字体については

      capital letter　　　　boldface type　　　Greek letter
      italic type　　　　　roman type

それらの具体的な使用は

      cis, trans　　　　　　constant
      crystal plane and direction
      crystallographic groups and space group
      fraction　　　　　　function　　　　　Latin term
      subscript　　　　　　symbols
      symmetry elements and point group
      variable　　　　　　vector

のほか，ラテン語とその略語が

      e.g.　　　etc.　　　i.e.　　　versus, vs.　　　via

の項目で説明されている．

   3.3 終止符，コンマ，ハイフン，かっこなどの形式については

      bracket　　　　　　centered dot　　　　colon
      comma　　　　　　　em dash　　　　　　en dash
      hyphen　　　　　　　parentheses　　　　period
      prefex　　　　　　　quotation mark　　　semicolon
      slash　　　　　　　　syllabication

の項目で，それぞれに使用方法が説明してある．

3.4 品詞一般については

  adjective   adverb   noun

  comparative and superlative adjectives

  comparative and superlative adverbs

の項目があり，文法一般については

  active voice, passive voice   double negative

  past tense, present tense

  restrictive clause, nonrestrictive clause

  singlar and plural

の項目がある．

3.5 動詞に単数形を用いるべきか，複数形を用いるべきかは singlar and plural の項目のほか，各所で扱われている．例えば，

  集合名詞では   majority   series   set   variety

  不定代名詞では   either   most   much   neither

          none   some

3.6 冠詞に関する注意は項目 a, an と the にまとめられている．

3.7 接続詞についても，多くの項目があるが，特に次の解説は重要である．

  and   because   but   or

3.8 接頭辞と接尾辞（連結形とよばれるものを含む）には，それぞれの項目 prefix, suffix がある．特に前者では，米式で例外的にハイフンを使用することがある接頭辞をまとめて示し，それぞれに独立した項目を設けて，解説した．接尾辞では，fold, 1ike に個別の項目がある．

3.9 まぎらわしく，使用に注意すべき語も取り上げてある．これには，

  affect, effect   complex, complicated

  find, discover   get, obtain   similar, like

のように，一つにまとめてあるものと

  commomly   in general   generally   ordinarily   usually

  part   portion   proportion

  partly, partially

のように，個別の項目として扱ってあるものがある．

 これらのほかにも，執筆に役立つと思われる数多くの項目が設けられている

ので，本文および索引を繰ってみられたい．

3.10 略語，記号の説明は abbreviation の項目に総括されている．広く普及しているものは裏見返しの別表に示してある．物理量の記号は Greek letter, italic type, physical quantities, 付表 1–12 を，それらの単位は units の項目を参照されたい．

3.11 数や数式にかかわる項目としては

comma	equation	fold
fraction	function	italic type
number, numeral	ratio	roman type
slash	symbols	units

がある．

4 索引の主体は日本語とし，配列は 50 音順による．

4.1 事項編と表現編の二つから成り立つ．

4.2 事項編には主見出しと副見出しが設けてある．一例をあげると，主見出しの「イタリック体」には，次の副見出しが見いだされ，参照先の項目が示されている．

```
イタリック体
  学名     capital letter (13)；italic type (14)
  化合物名の補助記号    abbreviation (11)；
      capital letter (1b)；italyc type (4)
  関数の記号    function (2)；italic type (6)
  強調のための使用    italic type (12)
  元素記号による位置記号    italic type (4)
  最初の定義での使用    italic type (13)；quota-
      tion mark (3)
  雑誌巻数    italic type (15)；references and
      notes (7)
  雑誌名，書籍名    italic type (15)；references
      and notes (7, 9)
  軸の記述    axis；italic type (7)
  物理量の記号    italic type (2)；physical quan-
      tities (1, 2)；subscript (1)
  ベクトルは太字の    boldface type (3)；vector
  ベクトル成分    italic type (8)
  変数の記述    variable (1–3)
```

5 本書でしばしば対比される The ACS Style Guide は，本文中では ACS と略記する．

例文作成の参考とした文献

例文は次の英米の単行本，総説誌ならびに論文誌を参考にし，本書に述べたスタイルに従って作成した．

単行本
◇P. Atkins, *Concepts in Physical Chemistry*, W. H. Freeman, New York, **1995**.
◇J. W. Baker, *Hyperconjugation*, Oxford University Press, London, **1952**.
◇G. M. Barrow, *Physical Chemistry*, McGraw-Hill, New York, **1961**.
◇F. Basolo, R. G. Pearson, *Mechanisms of Inorganic Reactions, A Study of Metal Complexes in Solution*, Wiley & Sons, New York, **1958**.
◇A. N. Campbell, N. O. Smith, *The Phase Rule and its Applications*, 9th ed., Dover Publications, New York, **1951**.
◇M. C. Day, Jr., J. Selbin, *Theoretical Inorganic Chemistry*, 2nd ed., Reinhold, New York, **1962**.
◇M. J. S. Dewar, *The Electronic Theory of Organic Chemistry*, Oxford University Press, Oxford, **1949**.
◇B. E. Douglas, D. H. McDaniel, J. J. Alexander, *Concepts and Models of Inorganic Chemistry*, 3rd ed., Wiley & Sons, New York, **1994**.
◇A. B. Ellis, M. J. Geselbracht, B. J. Johnson, G. C. Lisensky, W. R. Robinson, *Teaching General Chemistry, A Materials Science Companion*, ACS, Washington, DC, **1993**.
◇R. C. Evans, *An Introduction to Crystal Chemistry*, 2nd ed., Cambridege University Press, London, **1966**.
◇A. A. Frost, R. G. Pearson, *Kinetics and Mechanism*, 2nd ed., Wiley & Sons, New York, **1961**.
◇B. S. Furniss, *Vogel's Textbook of Practical Organic Chemistry*, 4th ed., Longman, London, **1978**.
◇K. B. Harvey, G. B. Porter, *Introduction to Physical Inorganic Chemistry*, Addison-Wesley, Reading, MA, **1963**.
◇J. E. Huheey, E. A. Keiter, R. L. Keiter, *Inorganic Chemistry, Principles of Structure and Reactivity*, 4th ed., HarperCollins, New York, **1993**.
◇N. N. Greenwood, A. Earnshaw, *Chemistry of the Elements*, Pergamon Press, Oxford, **1984**.

◇*Classics in Coordination Chemistry*, Part 3, ed. by C. B. Kauffman, Dover Publications, New York, **1978**.
◇I. M. Klotz, R. M. Rosenberg, *Chemical Thermodynamics*, 5th ed., Wiley & Sons, New York, **1994**.
◇K. J. Laidler, *Chemical Kinetics*, 3rd ed., HarperCollins, New York, **1987**.
◇K. J. Laidler, J. H. Meiser, *Physical Chemistry*, 2nd ed., Houghton-Mifflin, Boston, MA, **1995**.
◇W. F. Luder, S. Zuffanti, *The Electronic Theory of Acids and Bases*, Dover Publications, New York, **1961**.
◇J. W. McBain, *Colloid Science*, D. C. Heath, Boston, MA, **1950**.
◇D. A. MacInnes, *The Principles of Electrochemistry*, Dover Publications, New York, **1961**.
◇A. E. Martell, M. Calvin, *Chemistry of the Metal Chelate Compounds*, Prentice Hall, New York, **1952**.
◇*Chemical Sciences in the Modern World*, ed. by S. H. Mauskopf, University of Pennsylvania Press, Philadelphia, PA, **1993**.
◇F. M. Menger, D. J. Goldsmith, L. Mandell, *Organic Chemistry*, W. A. Benjamin, Menlo Park, CA, **1972**.
◇R. T. Morrison, R. N. Boyd, *Organic Chemistry*, 5th ed., Allyn & Bacon, Newton, MA, **1987**.
◇L. Pauling, *General Chemistry*, Dover Publications, Mineola, NY, **1970**.
◇S. S. G. Phillips, R. J. P. Williams, *Inorganic Chemistry*, Oxford University Press, Oxford, **1966**.
◇J. E. Ricci, *The Phase Rule and Heterogeneous Equilibrium*, Dover Publications, New York, **1966**.
◇J. D. Roberts, M. C. Caserio, *Modern Organic Chemistry*, W. A. Benjamin, New York **1967**.
◇H. Rossotti, *Diverse Atoms*, Oxford University Press, Oxford, **1998**.
◇R. T. Sanderson, *Chemical Periodicity*, Reinhold, New York, **1960**.
◇J. C. Slater, *Introduction to Chemical Physics*, Dover Publications, New York, **1970**.
◇D. W. Smith, *Inorganic Substances*, Cambridge University Press, Cambridge **1990**.
◇A. R. West, *Basic Solid State Chemistry*, Wiley & Sons, Chichester, **1988**.
◇G. W. Wheland, *Advanced Organic Chemistry*, 2nd ed., Wiley & Sons, New York, **1957**.
◇E. A. Wood, *Crystals and Light, An Introduction to Optical Crystallography*, 2nd ed., Dover Publications, New York, **1977**.

雑誌

◇Accounts of Chemical Research
◇Chemical Society Reviews
◇Chemistry in Britain
◇The Journal of the American Chemical Society
◇Journal of the Chemical Society
◇The Jornal of Organic Chemistry
◇Quarterly Reviews, Chemical Society

A

a, an

　一つのまたは**一人の**の意味の**不定冠詞**である．

　(1)　初めて話題とした**普通名詞**の単数形には，不定冠詞 a または an をその前につける．同一の名詞を話題とすることが明確であれば，次からは**定冠詞** the をつける〔☞ the (1)〕．

> ▶ When white light is passed through a substance, light of certain wavelengths may be absorbed by the substance.

　(2)　その種類全体に通じる一般的なことを述べるときに，**…というもの**の意味で，a, an を用いる．

> ▶ A rate law is the empirically determined relation between the rate of a reaction and the concentrations of the species that occur in the overall chemical reaction.

同じ目的で，the をつけた単数形，the をつけない複数形が用いられる〔☞ the (3)〕．

　(3)　集合体を 1 単位として見る**集合名詞**にも，a をつける．一つの集団と見なすときには単数形として扱い，集団を構成する個々を意味するときには複数形として扱う〔☞ majority, series, variety〕．

　(4)　**不可算名詞**である**物質名詞**と**抽象名詞**には，一つを意味する不定冠詞はつけない．なお，一般的なことを述べるときに，これらには定冠詞もつけない〔☞ the (4)〕．

　(5)　子音の前では a，母音の前では an を用いる．一般に略語は個々の文字を発音するので，その最初の文字が母音で発音される場合には an をつける．

　　a uranium complex　　　　an hour
　　a nuclear magnetic resonance spectrometer
　　an NMR spectrometer　　　an yttrium compound

　(6)　**元素記号**の発音．

　　(a)　元素記号が元素を表す場合には，その元素名として読む．その発音によって，a または an を選ぶ．

　　a Au wire　(a gold wire と読む)

a, an

　　<u>a</u> Xe lamp　　（<u>a</u> xenon lamp と読む）

　　<u>a</u> S-containing compound　　（<u>a</u> sulfur-containing compound と読む）

　（b）　**同位体**は元素名，ハイフン，次いで**質量数**を記す場合と，左上に質量数を付した元素記号で表す場合がある．後者の元素記号は文字として読み，次いで質量数を読む．それに応じて，a または an を選ぶ．

　　<u>a</u> hydrogen-3 isotope　　<u>an</u> ^{3}H isotope　　（aitch three と読む）

　　<u>a</u> carbon-14 isotope　　<u>a</u> ^{14}C isotope　　（c fourteen と読む）

　　<u>a</u> nitrogen-15 isotope　　<u>an</u> ^{15}N isotope　　（en fifteen と読む）

元素記号以外の記号の発音に関しては，次の abbreviation を見よ．

abbreviation

物理量の記号は付表 1-12 を，測定単位の記号は units を見よ．

　（1）　別表に示す**略語，記号**は広く普及しているので，本文中での使用にあたって，定義する必要はない．物理量の記号ではない mp や fw と数値の間には，等号＝を用いない．引用文献に用いる略語については，references and notes（1）を見よ．

　（2）　表題で略語，記号を用いることは避ける．

　（3）　概要，本文それぞれで，一度しか使用しない用語は略記しない．

　（4）　概要で定義した略語，記号も，本文で改めて定義する．

　（5）　別表にない用語を，概要や本文で繰り返し使用する場合は，その語を最初に用いたときに，それに続いて略語，記号をかっこ（　）に入れて定義する．その際，既に広く使用されている略語，記号があれば，これに従う．

　　atomic absorption spectroscopy（AAS）

　　Fourier transform infrared spectroscopy（FTIRS）

　（6）　終止符の使用は慣習に従う．別表に見るように，終止符を用いる例は多くない．atomic（at.），number（no. および No.）は，前置詞 at，形容詞または副詞 no と区別するために終止符を用いる．

　　ラテン語の略記には終止符が用いられる〔☞ Latin term（2）〕．

　（7）　数式に関係する次の略語も，最初に使用したときに定義する．終止符を使用するものと使用しないものがあることに留意せよ．

　　estimated standard deviation（e.s.d.）　　left-hand side（lhs）

　　ordinary differential equation（o.d.e.）　　subject to（s.t.）

　　root-mean-square deviation（rmsd）　　with respect to（w.r.t.）

(8) 必要ならば,著者が新たに略語,記号を定義してよいが,
　(a) 単位の記号とは異なるものを選ぶ.
　(b) 元素や基の記号とは異なるものを選ぶ.
　(c) 読者にとって理解しやすいもの,混乱のないものを選ぶ.
(9) 表の各列の冒頭は,記号／単位であることが望ましい.物理量ではない記号を表で使用するときには,表の脚注で定義する.また,表でのみ使用する略語についても同様に扱う.
(10) 略語,記号に,不定冠詞 a を用いるか,an を用いるかは,その発音の仕方による.元素記号単独の発音は次の規則に従わない〔☞ a, an (6)〕.
　(a) 通常の略語,記号では個々の文字として読む.
　(b) **頭字語**と呼ばれるものは,まとまった一つの語として読む.
　highest occupied molecular orbital(HOMO)
　lowest unoccupied molecular orbital(LUMO)
(11) 次の**有機原子団**の記号を定義する必要はない.ただし,化学式や化学構造式でのみ使用し,本文中での使用は避ける.

acetyl	Ac	aryl	Ar	benzoyl	Bz
butyl	Bu	ethyl	Et	methyl	Me
phenyl	Ph	propyl	Pr	alkyl	R

ローマン体の iso およびイタリック体の n-, sec-, $tert$- は,記号の前では,イタリック体の i-, n-, sec-, t- とする.たとえば,isobutyl は i-Bu,n-butyl は n-Bu とする.

(12) 略語,記号の複数形については,次の三通りの扱いがある.
　(a) 大文字だけからなるとき,および大文字でおわるときは小文字の s だけをつける.
　PAHs (PAH は polycyclic aromatic hydrocarbon の略語)
　pHs (pH は negative logarithm of hydrogen ion concentration の記号)
　(b) 大文字一字のとき,小文字からなるとき,上付きあるいは下付き文字でおわるとき,およびイタリック体でおわるときは 's をつける.
　oxygen の元素記号 O の場合,s だけをつけたのでは Os と区別できない.
　O's　　cmc's　　pK_a's　　P's
ただし,P's よりは P values の表現が望ましい.
　(c) eq, no., ref の複数形は eqs, nos., refs とする.

(12) 略語，記号を定義する例文については，subscript（3）も見よ．
- ▶ A_i indicates an atom A at an interstitial site, V_B a vacancy at a B site.
- ▶ K, S_A, N, and I stand for the crystalline, smectic A, nematic, and isotropic liquid phases, respectively.
- ▶ CB denotes the cyanobiphenyl group and 2MB the 2-methylbutyl group.

about, around
(1) **およその**意味の副詞として，口語では around が用いられるが，about の方が好ましい．次の例は望ましいものではない．
- ▶ Finding new and more efficient materials is important because the substances currently in use require fields of around several thousand volts per centimeter to produce a sufficient change in the refractive index.

(2) **…に関して，…について**の意味の前置詞として用いられる．
- ▶ Intramolecular hydrogen bonds in five- and seven-membered rings are also common, but we do not have data at this point about competition between these bonds and intermolecular hydrogen bonds.

類語 concerning, regarding, respecting, relating と意味の相違なく交換できる．

above
前置詞または副詞として用いるのが本来である．したがって，the paragraph above や in reference to the material above の方が，形容詞的に用いた the above paragraph や in reference to the above material よりは望ましい．
- ▶ In the nitroaniline examples above, the proton donors and acceptors were on the same molecule, but they should hydrogen bond to one another in the same way even if they are on different molecules.

according to
(1) **…に従って**の意味で用いられる．
- ▶ 5,10-Dihydro-5-methylphenazine was prepared under pure nitrogen according to the literature.
- ▶ Reduction of nitrobenzene in methanolic or ethanolic sodium hydroxide solution with zinc powder leads to azobenzene or hydrazobenzene according to the proportion of zinc powder employed.

(2)　**…によれば**の意味で用いられる．
▶ According to Brønsted's definition the species HA, H_3O^+, and BH^+ are acids, while B, OH^-, and A^- are bases.

account
(1)　**…の理由を説明する**の意味の account for の使用が多い．
▶ We can account for the actual geometry of water by postulating that the various orbitals of oxygen can mix, or hybridize.

(2)　take into account は**考慮する**の意味である．
▶ The vast majority of photochemical syntheses have been conducted in the liquid phase; hence, the apparatus assembly for a photochemical reaction must take into account the light transmission characteristics of the material from which the reaction vessels are made.

(3)　because of の同義語として on account of が用いられる．
▶ On account of the strongly electronegative character of oxygen many oxides AO_2 are primarily ionic and have symmetrical structures of the fluorite, rutile and β-cristobalite types composed of A^{4+} and O^{2-} ions.

active voice, passive voice
(1)　より簡明に述べるためには，**受動態**よりは**能動態**を用いる．
(2)　二つの文を一つにまとめた**分詞構文**においては，一般に分詞の意味上の主語は主文の主語に一致すべきである．この見地から，能動態，受動態の選択に注意を払う必要がある．次の例文の分詞の意味上の主語は we である．
▶ Dealing with an octahedral complex, we shall assume the presence of six ligands octahedrally coordinated about the metal atom.

もし，the presence of six ligands ないし six ligands を主文の主語とした受動態を選択すると，その分詞は主文の主語との文法的結合を失う〔☞ comma (5)〕．

adjective
(1)　**形容詞**のみのときは名詞の前に置く．複数個の形容詞を並べたとき，その順序を変えても意味を失わないならば，形容詞の間をコンマで区切る〔☞ comma (4)〕．句や節を伴うときは，名詞の後におく．次の例文はそのいずれをも含む．
▶ The product is a bright orange solid insoluble in water but soluble in some organic solvents.

(2) 他動詞, 自動詞の**現在分詞**は, 意味のうえで能動態を表す形容詞の働きをする. 前者は主に目的語と組み合せた形で使用される〔☞ hyphen (5) c〕.
- ▶ When a metal ion combines with an electron donor, the resulting substance is said to be a complex, or coordination compound.
- ▶ A polymorph is a solid crystalline phase of a given compound resulting from the possibility of at least two different arrangements of the molecules of that compound in the solid state.

(3) 他動詞の**過去分詞**は受動態を表す形容詞の働きをする. 自動詞, 他動詞の両方に使える動詞ならば, その意味によって現在分詞（能動態）と過去分詞（受動態）を使い分ける. すなわち, increasing は**次第に増える, 増大する**の意味, increased は**増加された**, decreased は**減少された**の意味となる.
- ▶ Polarizability increases with increasing size and charge for anions.
- ▶ Lithium is a metal: that is, it has high electrical conductivity which decreases with increased temperature.
- ▶ Increasing covalency from fluoride to iodide is expected, and decreased solubility in water is observed.

次の例文のように, increasing, increased は, なくても差し支えないことがある.
- ▶ Arrhenius showed that the rate constant increases in an exponential manner with the temperature.

(4) 形容詞は以下のように特定の前置詞を伴った形で用いられることが多いから, 辞典を参照することが望ましい.

attributable to	capable of
characteristic of	dependent on/upon
filled with	identical to/with
indicative of	parallel to/with
proportional to	subject to

adverb

副詞は文中における位置に定まった基準はなく, 強調したい位置においてよい. only は例外で, 修飾する語句の前または後におく.

形容詞と同形のものもあるが,

(1) 形容詞に接尾辞 -ly をつけた副詞が多い. 語尾が -ic の形容詞には

-ally をつけて副詞とする．early は例外的に形容詞でも副詞でもある．

approximate → approximately	rough → roughly
probable → probably	reversible → reversibly
quantitative → quantitatively	careful → carefully
interesting → interestingly	repeated → repeatedly
analytical → analytically	catalytic → catalytically

(2) 形の上では関連していても形容詞と副詞で意味が異なることがある．

(a) bare と barely 前者は**露出した，ありのままの**の意味の形容詞であり，後者は**わずかに，ほとんど…ない**の意味の副詞である．

(b) hard と hardly 前者は形容詞としては**固い，難しい**，副詞としては**熱心に**の意味がある．後者は余裕がないことを表す**ほとんど…でない**の意味で使用される副詞である．

barely は hardly よりも否定的な意味が弱いとされる〔☞ hardly〕．

(3) 名詞や形容詞に，**方向，方法**を意味する接尾辞 wise をつけた副詞もある．

anticlockwise	anywise	clockwise
dropwise	likewise	otherwise
pairwise	stepwise	

affect, effect

前者は**影響を及ぼす，作用する**の意味で用いられる動詞である．

▶ It has long been known that the rate of a chemical reaction may be greatly affected by atoms or groups which are close to the reactive centers of the reagent molecules, but which do not appear to be directly involved in the reaction itself.

▶ The electron distribution in one atom will affect and be affected by the electron distributions in the surrounding atoms.

間違えられやすい語に effect があるが，こちらは**結果として変化を生じる，果たす，遂げる**の意味で，両者は同義語ではない．

▶ The hydroxy group in tertiary alcohols is most readily replaced by chlorine, and this is effected by simply allowing the alcohol to react with concentrated hydrochloric acid at room temperature.

▶ Recrystallization was effected from a minimum amount of acetonitrile.

なお，effect of, effect on の形で，名詞としても使用される．
- ▶ The effect of substituents on the acidity of phenols is similar to their effect on carboxylic acids.

allow
(1) **可能とする**の意味で用いられる．
- ▶ The variation method not only allows reliable calculations to be made on complex atoms, but it also provides considerable insight into the nature of atoms and of chemical bonds.

(2) **見込む**の意味で使用される．
- ▶ A liquid is contained in a glass tube. The tube may be completely filled and allow space for expansion when the liquid becomes warm.

(3) **放置する，放冷する**の表現に be allowed to ... が用いられる．
- ▶ The solution is allowed to stand for crystallization of the product.
- ▶ The furnace was turned off and allowed to cool to room temperature.

後者の意味では be cooled on standing の表現もある．

alpha particle (α particle)
α 粒子の表記は α particle とし，alpha particle とはしない．なお，表題や文頭では，X-ray の r とは異なり，α Particle と p を大文字にする．

also
as well, too とともに肯定的に用いる〔☞ as well, too〕．一般に also は動詞の前におく．ただし，be 動詞の場合はその後におく．
- ▶ Somewhat surprising, hydrogen can also form σ bonds to transition metals, provided that some other group is also present.
- ▶ The intermolecular factors that are important in determining how a gas deviates from ideal behavior are also the quantities that govern the critical-state constants.

although
…であるが，しかしの意味の接続詞である〔☞ comma (7)〕．

among
…の間のの意味の前置詞で，三つ以上のものを対象とする．
- ▶ The remarkable resemblances among the lanthanides bear witness to the overwhelming influence of identical charge and similar size in these species.

▶ The elementary processes which are the units out of which actual reactions are built include unimolecular, bimolecular, and termolecular processes in which reaction occurs either spontaneously, at a collision between two molecules, or at a collision among three molecules, respectively.

amount
量を扱うときに用いる．数を扱うときには number を用いる．
(1)　amount of の後には名詞の単数形が続く．
▶ The amount of dioxygen in the atmosphere has probably been roughly constant for the last 500 million years.
(2)　複数形を用いた in small amounts は**一度に少しずつ**の意味である．

analytical data
測定結果の記載の仕方は IR spectroscopy, mass spectroscopy, melting and boiling points, NMR spectroscopy, quantitative analysis, specific rotation, UV–vis spectroscopy, X-ray diffractometry それぞれに挙げた例を見よ．

and
等位接続詞の一つで，連結されるものは，文法上，対等の関係にあることを要する．例えば，名詞と名詞であるのみならず，方法と方法，機器と機器など，同じ性格をもつことが必要である．
(1)　連結された主語が別個のものであれば，動詞は複数形にする．
　列挙するときには，A, B, C, and D や and so forth, and so on, and other things が用いられる〔☞ etc.〕．
▶ Ferromagnetism, antiferromagnetism, and ferrimagnetism are relatively rare phenomena in complexes and are of importance only in special cases.
(2)　2個，3個の名詞が一つの概念を表すときは，動詞は単数形にする．この場合，冠詞をつけるのは最初の名詞だけとする．
▶ The periodicity and fundamental importance of the size property has been recognized for a very long time.
▶ Oxidation and reduction of ligands due to their interaction with their central atoms has also received some attention.
attention は注意，注目，配慮の意では不可算名詞であるから，この some は特定されない量を表す〔☞ some (2)〕．
▶ The isolation, purification, and analysis of crystalline compounds is a

significant and widely used method of proving the existence of chelates.

(3) 名詞と名詞に限らず，文法上，対等の関係にある形容詞と形容詞，副詞と副詞，動詞と動詞，節と節を連結するのに用いる．形容詞を並べるときのコンマの使用については，comma (4) を見よ．二番目の例文のように，主語を共通にする二つの部分だけからなる叙述において，and の前をコンマで区切らないように注意する〔☞ comma (6)〕．

▶ The most important form of antimony is black, shiny, brittle, and fairy low melting.

▶ Silicon is the second most common element in the earth's surface and always occurs in combination with the commonest one, oxygen.

(4) 複数名の所有格では and で姓を連結し，最後の姓にだけ 's をつける．

Bausch and Lomb's equipment

ここで，equipment は不可算名詞であるから，不定冠詞は用いない．

(5) and の代わりに plus または＋を使用しない．

(6) and の代わりにスラッシュを使用しない．hot and cold を hot/cold と記すのは不可である．

and/or

and または or, both または either の意味の接続詞である．

▶ Every organic reaction involves the making and/or the breaking of chemical bonds.

and, or のいずれかで十分ではないか，検討してから使用するのが望ましい．

another

an other が一体化した形容詞または代名詞である．

(1) 可算名詞の単数形の前において**別の，もう一つ**の意味に用いる．

▶ There is an increase in acid hydrolysis at the lower pH due to acid catalysis, and another increase at the higher pH due to base catalysis.

(2) one another と each other を区別する必要はない．

▶ The addition reactions of acetylene resemble closely those of ethylene, since the two π bonds in it are perpendicular to one another and are, therefore, electronically independent.

anti-

(1) …**の反対の**の意味の接頭辞で，一般にハイフンを用いない．

antibonding　　　anticorrosion　　　antioxidant
antiparallel　　　antistatic　　　antisymmetric

(2) inflammatory, overshooting は，それぞれ接頭辞（in, over）を含む．これらに anti をつける場合には，既に含まれる接頭辞の綴りに関係なく，ハイフンを必要とする〔☞ prefix (2)〕．

anti-inflammatory　　　　　anti-overshooting

apparently
どうも…らしいまたは**明白に**の意味の副詞である．

▶ This new class of clathrate formers exhibits a sharp selectivity for forming crystals containing certain organic molecules, the selectivity being based, apparently, on the shape rather than on the molecular volume occupied by the organic moiety.

ACS は，it appears that や it would appear that よりは，apparently を使用するのが望ましいとする．

apply
(1) 他動詞で**適用する**，**応用する**，自動詞で**適用される**，**適合する**を意味する．

▶ It is quite another matter to see how this symmetry may apply to an actual crystal.

▶ A variety of techniques has been applied for thermal transitions in refractory metals.

▶ A similar scheme could obviously apply when the cataylst is an acid like tin-(IV) chloride.

(2) **…に当てる**の意味で使用される．

▶ When a powerful flash created by a discharge of 2,000–4,000 J through an inert gas is applied to a photochemically responsive system, very high momentary concentrations of atoms and free radicals are produced.

aqueous
水のような，**水を含む**の意味の形容詞である．aqueous ammonia（aqueous ammonia solution も見受けられる）はアンモニア水，aqueous M^+ は水溶液中の M^+，aqueous hydrochloric acid または aqueous solution of HCl は塩酸水溶液をさす．記号はそれぞれ，$NH_3(aq)$，$M^+(aq)$，$HCl(aq)$ となる．

are, is
（1）　論文には，口語である縮約形 aren't と isn't は用いない．
（2）　ACS は are found to be, are known to be よりは are を, has been shown よりは is を用いるのが望ましいとする．

arrows in reactions
化学反応に用いる**矢印**には次の種類がある．
（1）　反応には矢印→を用いる．　　$2H_2 + O_2 \rightarrow 2H_2O$
（2）　可逆反応には，　$CH_3COOH + C_2H_5OH \rightleftarrows CH_3COOC_2H_5 + H_2O$
（3）　平衡状態には，　$CH_3COOH \rightleftharpoons CH_3COO^- + H^+$
なお，↔は反応ではなく共鳴を，−は一電子移行を表す記号である．

as
（1）　**…として**の意の前置詞として，主として文頭で用いられる．

　　as an example　　　　　　　as a specific example of ...
　　as a result　　　　　　　　as a first approximation

（2）　**…ように**の意味の接続詞としても，主として文頭で用いられる．

　　as we pointed out above　　as the name implies
　　as is frequently the case　as already stated
　　as expected　　　　　　　　as mentioned in Section 1

（3）　**…すると**の意味の接続詞として用いられる．

▶ As particles become smaller, the properties of surface molecules become increasingly significant; surface chemistry is of fundamental importance in colloid science.

（4）　**…だから，…ので**での意味の接続詞としても使用される．as の用法が多いため，不明確になりやすいので，この意味での使用を避けることが好ましいとする考えがあり，since や because に比べると使用例は少ない．

▶ As silver is stable in damp air, it was used for coinage and for jewelry and small artifacts.

as for, as to
（1）　**…に関する限り**の意味では両者とも区別なく用いられる．

▶ Where effects are essentially inductive, as for all meta- and some para-substituents, σ values seem to be true constants, leading to good linear relations for many different reactions and equilibria.

(2) …について，…に比例しての意味では as to だけが用いられる．
- ▶ Whereas there is no consensus of opinions as to the cause of overvoltage, its existence helps to explain quite a number of electrochemical phenomena.

as well
そのうえの意味の副詞として，文の最後におかれる．
- ▶ The hydrogen-bonding relationships derived for nitroanilines are useful for understanding the hydrogen-bonding properties of other molecules as well.

as well as
…だけでなく，…と同様にの意味の接続詞として用いられる．A and B とは違って，A に重点がおかれる．動詞がこれに続く場合には，A が単数形か複数形かによって，動詞は単数形か複数形かが定まる．
- ▶ The majority of prototropic reactions can be catalyzed by acids as well as by bases.
- ▶ There is a wide divergence in the chemistry of the elements involved as well as in the reducing agents and conditions used.

assume
仮定するのほかに，**執る**，**帯びる**の意味でも使用される．
- ▶ Iron can assume the oxidation states $+2$, $+3$, and $+6$, the last being rare, and represented by only a few compounds, such as potassium ferrate, K_2FeO_4.
- ▶ If a very dilute solution of surfactant molecules is dropped on the surface of pure water, and the solvent evaporates or dissolves, the surface-active molecules are restricted to the interface with their polar head groups bound to the aqueous subphase and their fatty tails assuming various orientations relative to the surface plane, depending upon the available area per molecule.

asterisk
論文の通信連絡にあたる著者の姓の右上に＊をつける．＊は励起状態を表す記号でもある．

atomic orbital
原子軌道は AO と略記する．
(1) 主量子数 $n = 1, 2, 3 ...$ に対応する**電子殻**の記号にはローマ体の大

文字を用いる．

    K   L   M   N   O   P   Q

(2) **副殻**と**軌道**の記号にはローマン体の小文字を用いる．

    1s electron   3d orbital sp^2 hybrid orbital

(3) p や d 軌道の軸は，右下にイタリック体で付記する．すなわち，

    p_x p_y p_z d_{xy} d_{yz} d_{xz} d_{z^2} $d_{x^2-y^2}$

atoms and molecules

原子，**分子**に用いる物理量の記号と SI 単位については付表 1 を見よ．

auto-

(1) **自身の**，**自動の**の意味の連結形で，ハイフンは使用しない．

    autocatalysis   autoclave   autoignition

    autoionization   autolysis   autoradiography

(2) 次の語の綴りは例外的である．

    autoxidation

axis

軸はイタリック体の小文字を用いて，*a* axis のように記し，ハイフンは使用しない．

based on

based は…**に基礎をおく**の意味の形容詞であるから，be 動詞か，修飾すべき名詞の後に位置するのが正しい用法である〔☞ adjective (1)〕．

▶ Photochromic systems are <u>based on</u> the ability of certain compounds to undergo reversible molecular rearrangements under the influence of light which results in a color change.

▶ Experimental approaches to inorganic syntheses are often limited by conventional ideas <u>based on</u> the limits imposed by hydrolysis occurring in the narrow window of acidities and basicities available in the common solvent water or by the redox limits of that solvent.

次の例文のように，based on を文頭におくのは正しい用法ではないから，on the basis of と取り替える．

▶ <u>Based on</u> theoretical considerations first given by Evans and Polanyi, Semenov has suggested that for exothermic abstraction and addition reactions of atoms and small radicals, the following approximate equation may be used: ....

basis, basic

(1)　basis は**根拠**，**基礎**の意味の名詞として用いる．

▶ Lewis chose four familiar experimental criteria as the <u>basis</u> of his definitions of acids and bases.

▶ On this <u>basis</u> it can be expected that a small amount of free liquid will pull itself together to form a more or less spherical drop.

(2)　…**に基づいて**の意味の on the basis of として用いる．

▶ <u>On the basis of</u> the collision theory these reactions require an assignment of a probability factor of 10^{-4} or 10^{-5}.

▶ Since so many factors are involved, the development of ion-selective electrodes is necessarily done <u>on the basis</u> of a good deal of empiricism.

(3)　basic は**基本的な**を意味する形容詞として用いられる．次の例文では，補語を強調するために，文頭に出している．

▶ Basic to the understanding of photochemical processes is an appreciation of the energy acquired by a molecule as a result of the absorption of light.

because

(1) …だからの意味で，since や as と同じように用いられる接続詞である．ACS は because の使用が望ましいとする．

▶ Because the infrared spectrum of a substance is so uniquely characteristic, it is often referred to as a fingerprint spectrum.

▶ HCl is an acid by virtue of the fact that it can donate a hydrogen ion, not because it might have already done so in an aqueous solution.

▶ These substitution reactions are called electrophilic because, as we shall presently see, the species attacking the aromatic rings are electron deficient.

(2) …のためにの意味の because of の形で用いる．同じ意味の on account of, owing to より多く使用されている．

▶ Because of the low natural abundance of most heavy isotopes, these isotropic peaks are generally much less intense than the M^+ peak; just how much less intense depends upon which elements they are due to.

▶ The study of the hydrogen–oxygen reaction has attracted much attention, not only because of its technical importance but because of the simplest branched-chain reaction.

beta particle（β particle）

β 粒子の表記は β particle とし，beta particle とはしない．なお，表題中や文頭では，β Particle と p を大文字にする．

between

…の間でなどの意味で，二つを対象として用いる前置詞である．三つ以上のものを対象とするときには among を用いる〔☞ among〕．

▶ The reaction between hydrogen and iodine is known to take place at bimolecular collisions involving a single molecule of each kind.

▶ Between two oxygen atoms the forces are inappreciable when the distance between nuclei is greater than 3 Å.

それぞれが複数形であることは妨げない．

▶ Semiconductors fall between insulators and conductors in that the electrons are localized but with a small energy gap.

次の例では，三つ，四つに言及されているが，いずれも二つずつを対象とする表現をまとめたものとみなされる．
- ▶ The main features of interest in the rich and varied chemistry of mixed-valence complexes are the relationships between gross physical properties, molecular structure, and the extent of valence-electron delocalization.
- ▶ Other correlations between the melting point, boiling point, enthalpy of sublimation, and enthalpy of vaporization are also present in the same data set.
- ▶ One type of stabilizing interaction is between certain metal ions such as Ni^{2+}, Cu^{2+}, and Zn^{2+} and the ammonia molecule, which acts as ligand to form stable ammine complexes.

bi-
(1) **二つ**の意味の接頭辞で，一般にはハイフンを用いない．

biaxial	bicyclic	bidentate
biradical	birefringent	bivalent

(2) 接頭辞が重なるときにはハイフンを使用する〔☞ prefix (2)〕．

bi-univalent

boldface type
ゴシック体は次の場合に使用される．

(1) 論文の表題および experimental, results, discussion, references, figure, scheme, table の見出しに用いる．

(2) **化合物番号**には，ゴシック体のアラビア数字を用いる．

 (a) 正式名の後には，かっこ（ ）に入れた番号を付する．

(1,2,3-η-2-butenyl)tricarbonylcobalt(I) (**1**)

2,3-quinolinediacetic acid (**2**)

poly[formaldehyde-*alt*-bis(ethylene oxide)] (**3**)

 (b) 化合物の正式名が記載されていないとき，構造を図示したとき，あるいは既に番号が与えられているときは，かっこなしの番号でこれを記す．

keto alcohol **4**

compounds **2**, **5**–**8**, and **10**

ACS では上記の **5**–**8** の二分ダッシュも太くしている．

(3) **ベクトル**（付表 3 の *B*, *D*, *H*, *M* ほか），**テンソル**，**行列**（マトリッ

クス）などには，イタリックゴシック体を用いる．

　（4）　論文誌，単行本，特許など，引用文献の発行年をゴシック体とする〔☞ references and notes〕．

bond, bonding orbital

本文中の化学式では，**結合**が考察の対象になっていない限り，**単結合**を二分ダッシュで表すことはしない．

　（1）　考察に必要ならば，単結合には－（ハイフンではなく，二分ダッシュ）を，**二重結合**には＝を，**三重結合**には≡を用いる．**水素結合**や**会合**を表すには…を用いる．

　（2）　**軌道**や結合に用いる π や σ はギリシャ文字で表し，pi や sigma とはしない．この際，ハイフンは用いない．

　　　π orbital　　　　　　σ bond

book

単行本の引用例は references and notes（9）を見よ．

both ... and ...

両者のを意味する both は A and B を強調するだけで，両者の間に軽い重いはない．次の例文中の spectroscopy は複数形のない抽象名詞である．

▶ Both infrared and Raman spectroscopy have selection rules based on the symmetry of the molecule.

A と B は名詞に限らないが，文法上，対等の関係にあることを要する．

▶ Titanium, unlike scandium, is both fairly common and widely used; indeed, it is the second commonest transition element after iron.

▶ Information in the literature on the ammines of nickel cyanide is both confused and contradictory.

▶ The fact that triethylammonium dichlorocuprate(I) can be both oxidized and reduced is an asset in the above reactions, but it is a handicap when the system is used merely as a solvent.

bracket

角かっこ（ブラケット）[ ]の主な用例を次に記す．山形 ⟨ ⟩ の用例は crystal plane and direction を見よ．

　（1）　関係式中の濃度の記載に用いる．

　　　$K_a = [H^+][A^-]/[HA]$　　　　　　$[Ca^{2+}] = 3 \times 10^{-2}$ M

ただし，本文中では $[Ca^{2+}]$ ではなく，calcium concentration と記す．
（2） 化学式や化合物名において，かっこ（ ）だけでは不十分な場合に，角かっこ［ ］を併用する．

$Na_2[Fe(CN)_5(NO)]$

3-bromo-1-[bis(ethoxycarbonyl)methyl]indole

（3） **橋かけ炭化水素**，**スピロ炭化水素**の名称の中の炭素原子数を示す数字は，角かっこ［ ］に入れて表示する．

bicyclo[3.2.1]octane

［ ］内の炭素鎖の長さを表す数字の区切りは終止符（ピリオド）とする．

（4） **縮合環化合物**の縮合位置を示す記号，番号は角かっこ［ ］に入れて表示する．

dibenzo[*f, h*]quinoline　　　naphtho[2, 1-*b*]furan

［ ］内の記号，番号の区切りはコンマとする．

（5） **同位体修飾化合物**　同位体で修飾した化合物には，**同位体置換化合物**と**同位体標識化合物**がある．この後者の化学式に角かっこ［ ］を用いる．同位体置換された化合物名については parentheses (4) を見よ．

（a）　形式上，特定の1種類の同位体置換化合物を同種類の無修飾化合物に加えたと見なされる**特定数標識化合物**では，核種の位置と数ともに定まっている．この場合は，核種の記号を［ ］の中に表した化学式を書く．

$H[^{36}Cl]$　　　　　$[^{13}C]O[^{17}O]$

（b）　特定数標識化合物の混合物と見なされる**特定位置標識化合物**で，核種の位置は規定されているが，数は必ずしも定まっていない場合には，核種の記号を角かっこ［ ］に入れて，化学式の先頭につける．

$[^{36}Cl]SOCl_2$　　　　　$[^2H]PH_3$

（6） **結晶軸**とその方向の記載に角かっこ［ ］を用いる〔☞ crystal plane and direction〕．

the [001] axis　　　the [101] direction

bring about

引き起こす，もたらすの意味の他動詞である．

▶An extremely small amount of a catalyst frequently <u>brings about</u> a considerable increase in the rate of a reaction.

but

　等位接続詞の一つで，文法上，対等の関係にあるものを連結する．…がの意では，nevertheless よりは意味が弱く，however よりは意味が強いとされる〔☞ comma (6), (11)〕．
　▶ The NMR spectra of alkanes are reasonably characteristic but difficult to interpret, because the chemical shifts between the various kinds of protons are rather small.

by

　一般に受動的表現において用いられる．**よって**を意味する by は動作主を，**用いて**を意味する with は測定方法や機器を表すとされていたが，近年は後者の場合にも by が用いられる〔☞ with〕．
　▶ The components were characterized by ^{1}H and ^{13}C NMR on a Bruker WM 250 spectrometer.
　この例文の後半は
　　by ^{1}H and ^{13}C NMR using a Bruker WM250 spectrometer.
　　by ^{1}H and ^{13}C NMR (Bruker WM250 spectrometer).
としてもよい．
　同じ名詞であっても，語尾が -graphy, -metry, -scopy である測定方法と，語尾が -graph, -meter, -scope である機器は対等の関係にはないから，並列しないように注意する．次の例は方法で統一されている．
　▶ The final product was purified by chromatography on alumina followed by recrystallization, and identified by X-ray methods.

by-

　副次的の意味の接頭辞で，化学用語はハイフンを用いた次の例に限られる．
　　by-product

calculated
calcd と略記する．これに対する語は found または observed で，後者は obsd と略記する．

cannot, can not
通常は cannot を使用する．論文には縮約形 can't は用いない．

capital letter
単語の頭字をどのような場合に**大文字**とするかを，次にまとめる．なお，表題には別の方式が用いられるから title を見よ．

(1) 文頭では頭字を大文字とする．

例外として

　(a) X-ray の場合とは異なって，α particle は α Particle，γ ray は γ Ray とする．

　(b) 化学物質名の前にハイフンを伴うイタリック体の位置記号 *o*, *m*, *p* や接頭辞 *sec*, *tert*, *cis*, *trans*, *sym*, *anti*, *erythro*, *threo* などは小文字のままとする．

　　cis-Azobenzene　　　　　　*o*-Dichlorobenzene

(2) 文頭では，化学物質名の一部ではないローマン体の cis, trans, syn, anti, erythro などは大文字とする．

　　Cis isomer　　　　　Trans hydroxy group

(3) 文頭では，ローマン体の元素記号が前についた名詞や形容詞は大文字とする．

　　N-Oxidation　　　　O-Substituted

(4) 文頭においては，かっこ ( )，[ ]，{ } の中に複雑な置換基が含まれているとき，かっこの内外を問わず，頭字だけを大文字とする．

　　4-{[4-(Dimethylamino)phenyl]azo}benzenesulfonic acid

　　N,N-Dimethyl-4-(phenylazo)benzenamine

(5) 文頭においては，接頭辞 poly に続くかっこの中に重合する化学種が示されているとき，poly の頭字 p だけを大文字とする．

　　poly(acrylic acid) は，Poly(acrylic acid) とする．

(6) 用語の一部となった固有名詞の頭字は常に大文字，これに続く名詞の頭字は常に小文字とする．

  Avogadro's number   Faraday constant
  Raman spectrum   Schiff base

ただし，Nobel Prize は例外とする．

(7) 固有名詞に由来する形容詞の頭字は常に大文字とする．

  Coulombic   Hamiltonian

(8) 番号で特定された figure，table，scheme は大文字とする．

  Figures 1–3   Table 2   Scheme 3

なお，page は大文字にはしない．

(9) 本から引用した特定の chapter，section，appendix は大文字とする．

  Chapter 1   Section 2.1   Appendix I

(10) 商標は大文字ではじめ，形容詞として用いて，適当な名詞をつける．表題中には商標を使用しないで，一般的な名称を用いる．

  Pyrex ではなく borosilicate glass
  Celite ではなく diatomaceous earth
  Butyl Cellosolve ではなく ethylene glycol monobutyl ether
  Nujol ではなく mineral oil
  Carbowax ではなく poly(ethylene glycol)

(11) 普通名詞である機器の名前には大文字は用いない．製造会社名は大文字ではじめる．番号や略号を伴う model は大文字にしない．一台であれば，不定冠詞を添える．

  a Perkin-Elmer model 240C mass spectrometer

(12) 元素記号はローマン体を用い，その頭字は大文字とする．

(13) 生物の学名は，イタリック体を用い，属名の頭字は大文字とし，種名はすべて小文字とする．後者は表題においても，大文字にはしない．species の略語（sp. や spp.）はローマン体の小文字とする．

  Bacillus subtilis   *Salmonella* sp.

ただし，属名を普通名詞として扱うとき，属名の複数形または形容詞，ならびに科名にはローマン体を用い，文頭，表題，見出し以外では，頭字も小文字とする．

  bacillus   bacilli   Leguminosae（マメ科）

caption

図表の説明のほか，書物の章，節などの**題目**をいう．

(1) 図の題目は図の番号を含み，簡潔に情報を提供する記述であることを要する．大文字ではじめ，終止符はつけるが，文章の形でないのが好ましい．

▶ **Figure 1.** Structure of the hexagonal and rhombohedral forms of graphite.

説明のための記号や，より具体的な説明にはいる前はコロンで区切る．それに続く説明の間はコンマで区切る．必要ならば，さらにセミコロンを用いて区切る〔☞ semicolon〕．

▶ **Figure 3.** Rate of reaction of ethylene chlorohydrin with base in mixtures of solvents: (a) water/dioxane, (b) water/ethanol, (c) water/methanol.

図面（scheme）の扱いは図に準ずる．ただし，題目は必ずしも必要としない．

(2) 表に添える題目は内容を簡潔に伝えるものとし，大文字ではじめるが，文章の形にはしない．本文を参照せずに理解でき，本文にない情報を含まないことが肝要である．図の題目とは異なり，終止符をつけない．

▶ **Table 1.** Crystal data for compounds **3** and **6**

centered dot

中黒・のこと．次の (1) と (2) の場合は，・の前後は詰める．

(1) 単位記号の積を表すのに用いる．**SI 誘導単位**を SI 基本単位で表すと，

enthalpy　　　　J　　　　$m^2 \cdot kg \cdot s^{-2}$
resistance　　　Ω　　　　$m^2 \cdot kg \cdot s^{-3} \cdot A^{-2}$
capacitance　　F　　　　$m^{-2} \cdot kg^{-1} \cdot s^4 \cdot A^2$

ただし，他の表現の仕方もある〔☞ units (12)〕．

(2) **付加化合物**，**水和物**の化学式に用いる．

$BF_3 \cdot 2H_2O$　　　　　　$(C_2H_5)_2O \cdot BF_3$
$3CdSO_4 \cdot 8H_2O$　　　　$Al_2(SO_4)_3 \cdot K_2SO_4 \cdot 24H_2O$

(3) **遊離基**の化学式において，・は不対電子を表す．

・H　　　・CH_3　　　・SH　　　・C_6H_5

ただし，右上に dot をつけてもよい．

H・　　　CH_3・　　　HS・　　　C_6H_5・

遊離基が電荷を帯びるとき，すなわち**陽イオンラジカル**，**陰イオンラジカル**の場合には右上に dot をつけ，続いて電荷を記する．

$R^{\cdot +}$　　　$R^{\cdot -}$　　　$C_3H_6^{\cdot +}$

質量分析では，・と正負の記号の順序が逆となることに注意せよ．
chemical element
element を見よ．
chemical equation
化学方程式（化学反応式）は短いものであれば，本文に含めてもよい．

(1) 必要があれば，改行して番号をつける．番号づけは数式と共通にしても，1, 2, 3 と I, II, III を用いて，別扱いにしてもよい．

(2) 物質の収支に必要な反応物，生成物につける化学量論係数と化学式との間は詰める．使用する矢印に関する説明は arrows in reactions を見よ．

$$H_2SO_4 + 2NaOH \rightarrow Na_2SO_4 + 2H_2O \qquad (1)$$
$$Ni + 4CO \rightleftarrows Ni(CO)_4 \qquad (2)$$
$$(COOH)_2 + 2C_2H_5OH \rightleftharpoons (COOC_2H_5)_2 + 2H_2O \qquad (3)$$

(3) 物質の状態を示すには，化学式の後に (s), (l), (g), (aq) をつける．これらと化学式との間は詰める．

$$Zn(s) + 2HCl(aq) \rightarrow ZnCl_2(aq) + H_2(g) \qquad (4)$$

(4) 反応条件や触媒を示すには矢印を長くして，その上下に小さい文字や記号で記入する．光のエネルギーは $h\nu$，熱エネルギーは Δ で表す．

(5) **核反応**，$^{14}N + {}^4He \rightarrow {}^{17}O + {}^1H$ は $^{14}N(\alpha, p)^{17}O$ と略記する．さらに例をあげると

$$^{11}Be + {}^2H \rightarrow {}^{12}Be + {}^1H \qquad\qquad ^{11}Be(d, p)^{12}Be$$
$$^{59}Co + {}^1n \rightarrow {}^{60}Co + \gamma \qquad\qquad ^{59}Co(n, \gamma)^{60}Co$$
$$^{23}Na + {}^1H \rightarrow {}^{21}Mg + 3{}^1n \qquad\qquad ^{23}Na(p, 3n)^{21}Mg$$

(6) 反応の型はローマン体の大文字と数字を組み合わせた記号で表す．

first-order nucleophilic substitution	S_N1
second-order radical nucleophilic substitution	$S_{RN}2$
first-order elimination	$E1$

chemical kinetics
化学反応速度論に用いる物理量の記号と SI 単位については付表 2 を見よ．
chemical name
化学物質名のうち，元素名は element を見よ．また，化合物命名法はその専門書を参照せよ．ここでは，一般的な注意事項のみを記する．

(1) 化合物名はローマン体を用い，普通名詞として扱う．

(2)　化学式にはローマン体を用いる．
　(3)　本文中に化学物質名と化学式を併用してもよいが，一つの物質名の中に，単語と記号を混用してはいけない．すなわち

　　　magnesium chloride hydroxide　　　　MgCl(OH)

のいずれを用いてもよいが，

　　　Mg chloride hydroxide　　　　　　　magnesium Cl(OH)

としてはいけない．

　(4)　二語からなる化学物質名を形容詞として用いるときには，ハイフンは使用しない〔☞ hyphen (6)〕．

　　　silver iodide precipitate　　　　　　acetic acid concentration

　(5)　化学物質名に接頭辞 non をつけるときにはハイフンを用いる．

　　　non-hydrogen atom　　　　　　　　non-alkane

　(6)　化学物質名に接尾辞 like をつけるときにはハイフンを用いる．

　　　adamantane-like　　　　　　　　　olefin-like

cis, trans

　(1)　一般にはローマン体を用いる．次の語との間にはハイフンを用いない．文頭においては，頭字を大文字にする．

　　　cis form　　　　　　　　　　　　Cis form
　　　trans effect　　　　　　　　　　　Trans effect

　(2)　ただし，化学物質名の一部をなすときには，イタリック体でハイフンを伴う．文頭においても，イタリック体の部分は小文字のままとする．

　　　cis-azobenzene　　　　　　　　　*cis*-Azobenzene

　〔☞ capital letter (1), (2) ; italic type (4)〕

citation in text

　(1)　本文中の**文献の引用**は，該当箇所の右上にその番号を示す．文末の場合は，終止符の後におく．一ヵ所で複数の文献を引用するときは，番号順に並べてコンマで区切る．連続した番号はダッシュで示す．

　　　in the literature[2,7-9]

　(2)　文献の著者が二人ならば，その名を and を用いて並べる．三人以上ならば，一番目の著者の名だけを書いて，et al. と続ける．

　　　Pearson[5]　　　　Kranz and Clark[9]　　　　Bryce et al.[11]

　(3)　一番目の著者は同じで，共著者が異なる複数の文献を引用するときは

Pauling and co-workers[1,2]　　Fischer and colleagues[5-8]

co-

(1)　**共通の，相互の，共同の**の意味の接頭辞で，一般に，oが重なる場合も含めて，ハイフンは用いない．

　　coauthor　　　　cooperation　　　coordination
　　coplanarity　　 copolymer　　　　coprecipitation

(2)　例外は

　　co-ion　　　　　co-oligomer　　　co-worker

colon

コロン：は主に導入の符号で，その後に1字分あける．動詞と目的語，補語の間，前置詞とその目的語の間を，コロンで区切らないように注意せよ．

(1)　前の節の例証，換言，要約，敷衍を後の節が行うとき，両者の間にコロンをおく．

▶ There is yet another most interesting group of boron compounds: the nitrides and their derivatives.

▶ Colloidal particles may be formed in two distinct ways: by subdivision of bulk material or by growth from molecular dimensions.

コロンの後，後続の語句の独立性が強ければ大文字で書き出す．

▶ Obvious defects to be introduced on the quasicrystalline model include: (1) Creation of vacant lattice sites. (2) Insertion of additional atoms at interstitial sites between the normal lattice points.

(2)　測定値の記載の際に使う．

^{1}H NMR (500 MHz, CDCl$_3$, δ): ....

IR （cm^{-1}）: ....

MS m/z (relative intensity): ....

UV （CH$_3$OH） λ_{max} (log ε): ....

Anal. Calcd for C$_{12}$H$_{20}$N$_2$O$_8$S: .... Found: ....

ここで，....は語や節の省略を意味し，四番目の．は終止符である．

(3)　図の説明において，より具体的な説明に入る前はコロンで，それらの説明の間はコンマ，必要ならば，さらにセミコロンで区切る．

▶ **Figure 1.** Molar solubilities in water–dioxane mixtures: (○), silver acetate; (△), silver sulfate; (□), barium iodate.

(4) **比**を表すには，コロンまたはスラッシュを用い，その前後は詰める．なお，**混合物**の成分の間には，スラッシュまたは二分ダッシュを用いて，コロンは使用しない．比率にはコロンまたはスラッシュを用いる．

 a hydrogen–oxygen (2:1) mixture
 a hydrogen–oxygen (2/1) mixture
 a hydrogen/oxygen (2:1) mixture
 a hydrogen/oxygen (2/1) mixture

comma

コンマは文中の切れ目を表す符号で，その後に一字分あける．動詞とその主語，目的語などの間を，コンマで区切らないように注意せよ．

(1) 語，句，節を列記するとき，次の二つのコンマの用い方が行われているが，前者が好ましいとされる．

　　A, B, and C　　　　A, B and C

(2) 文頭の however, therefore, thus, その他の導入の語や句など，文中の特定の部分を修飾するものではなく，文全体に関係するものはコンマで区切る．したがって，複数の節からなる文において，導入の語や句を使用する場合には，それが文全体に関係するか否かに注意を払う．

▶ Accordingly, the most important applications of EXAFS to date have come in the area of disordered materials and heterogeneous systems.

▶ Up to the present, structural studies on discontinuous transformations have seldom been correlated with thermodynamic considerations.

▶ Upon recrystallization, this impurity may dissolve in the boiling solvent and be partly adsorbed by the crystals as they separate upon cooling, yielding a colored product.

(3) 語句が文中に挿入されたときや，however, therefore などの副詞が挿入されたとき，前後をコンマで区切る．

▶ Barium, like strontium, is a very reactive metal. With nickel, however, it forms an alloy that is stable enough to be used in sparking plugs.

(4) 名詞の前に形容詞が列記されているとき，その順序を変えても意味が変わらないならば，形容詞をコンマで区切る．

▶ Bromine is a volatile, red, pungent liquid.

他方，形容詞の順序を変えると意味が変わる場合には，相互に区切らない．

    white basic carbonate
    important industrial polymer

(5) **分詞構文**は文頭と文尾いずれにあっても，主文との間をコンマで区切る．なお，分詞の意味上の主語が主文の主語とは異なる**独立分詞**には，active voice, passive voice (2) に述べた制約はない．独立分詞に使用できる分詞には，次のものがある．

concerning	considering	failing
given	judging	provided
providing	regarding	

次の二つは，特に数学的表現でよく使用される．

    assuming   taking

▶ Considering this fact, along with the observation that the reaction occurs with a decrease in volume, it is possible that both hydrogen bonding and hydrophobic forces are important in the dimerization of Rhodamine B in aqueous solution.

▶ Having more valence atomic orbitals than valence electrons, electron-deficient compounds present ambiguities in the use of the paired electron bond.

▶ Providing that the components are adequately volatile, gas-liquid chromatography (GLC) is perhaps the most powerful technique for the rapid and convenient analysis of the composition of mixtures of organic compounds.

▶ Provided that the correct growth conditions are established, it is possible to deposit the active layer as a perfect single crystal film in good registration with the substrate.

(6) 主語を共通にする二つの部分からなる叙述では，**等位接続詞**（and, but, for, nor, or, so, yet）の前にコンマを用いてはいけない〔☞ and (3)〕．

▶ Absorption and emission spectroscopy are the techniques most widely used and yield the most information about aggregation of complex organic molecules in aqueous solution.

▶ Anhydrous calcium chloride has a high water-absorption capacity but is not very rapid in its action; ample time must, therefore, be given for desiccation.

(7) **複文**：節の一つが主になっていて，それに他の節が従属した文，すなわち，主語と動詞の組み合わせが二つ以上あって，**従属接続詞**（although, as, because, if, since, unless, while など），関係詞，疑問詞で結ばれている文をいう．主節の前に**従属節**がある場合には，その後にコンマを用いる．

- Although major impurities in the commercial grades of acetone are methanol, acetic acid and water, the analytical reagent generally contains less than 0.1 percent of the organic impurities although the water content may be as high as 1 percent.
- Because skeletal components differ in density and crystallinity, they can be contaminated at different rates.
- Since these two natural radioactive decay series produce different isotopes, the relative atomic mass of lead can be stated only very imprecisely.

(8) 主節と従属節の順序が (7) とは逆の場合は，従属節が補足説明を加える**非制限的**なときにだけ，従属接続詞の前にコンマを用い，制限的なときにはコンマを用いない〔☞ restrictive clause, nonrestrictive clause〕．

- Complexes which contain the η^2-H_2 ligand are now referred to as nonclassical, while those in which the H–H bond has been severed are called classical.

(9) 文中にある非制限的な句や節の前後は，コンマで区切る．

- A specimen of granite, in which grains of three different species of matter can be seen, is obviously a mixture.

(10) such as, including で導入された句には，**制限的**な場合と，補足的説明にすぎない非制限的な場合とがある．制限的な場合はコンマを使用しない．

- A class of compounds which is most important in the life processes of aerobic organisms are the heme proteins and related substances such as the chlorophylls.

非制限的な場合は，(9) と同じくコンマを使用する．

- A very valuable feature of magnetic domains in optically transparent materials, such as the orthoferrites and garnets, is that they can be observed directly using the Faraday effect, i.e., the rotation of the plane of polarized light by the magnetic vector.

(11) **重文**：それぞれが独立することができる節からなる文，すなわち，主

語と動詞の複数の組合せが等位接続詞で結ばれている場合には，等位接続詞の前にコンマを用いる．comma (6) との相違に注意せよ．

▶ An acid is a species that tends to give up a proton, <u>and</u> a base is a species that tends to accept a proton.

▶ Glass is normally thought of as transparent, <u>but</u> if the edge of a sheet of ordinary window glass is examined, it appears dark green.

なお，前の例文における give up は**手放す**，**捨てる**の意味で，よく用いられる表現である．

　等位接続詞に続く節においても，コンマの使用は独立した文におけると同様に行う．

▶ The elements of the lanthanide and actinide series are all metals, and, on account of the fact that the differentiating electrons are so deeply buried in the electronic structure, they show great similarity in chemical properties.

(12)　文中の that is, namely, for example の後が，節ではなく，項目または一連の項目の場合には，これらの語の前後をコンマで区切り，各項目の後にもコンマを用いる．ラテン語に基づく i.e., e.g. を使用した場合のコンマの用法については，i.e. と e.g. それぞれを見よ．

▶ With rise of temperature more random or disordered structures, <u>that is</u>, <u>those of higher entropy</u>, become more favored.

(13)　項目に，かっこ（ ）の中の一連の項目の記載が付随するとき，かっこ内の項目にもコンマは使用される．なお，本文中の本来あるべきコンマも，忘れずにつける．

▶ Classical thermodynamics deals only with measurable properties of matter in bulk (e.g., pressure, temperature, volume, electromotive force, magnetic susceptibility, and heat capacity).

▶ The structure was confirmed with spectroscopy (^{1}H NMR, UV, and IR), high-resolution mass spectroscopy, and elemental analysis.

(14)　引用文の前にはコンマをおく〔☞ quotation mark〕．

▶ In the words of G. N. Lewis, who always acknowledged his debt to Werner, "We must consider Werner's theory of coordination numbers as the most important principle at present available for the classification of polar compounds."

ただし，完結した文として引用されていない場合には，コンマは用いない．
▶ Pauling has noted that Dickinson's work "has been found over twenty-five years of check by modern methods to be completely free from error."
(15)　五桁を越える数には，次の二通りの扱いがある．
　　(a)　小数点の左側三桁ごとに，コンマを使用する．
　　　　4378　　　　　　20,000　　　　　　390,582
　　(b)　単位がついているときには，四桁以上の数は避けた表現をとる．
　　　　44,000 kg → 4.4 × 10^4 kg　　　　20,000 mL → 20.0 L
(16)　引用文献番号については，citation in text を見よ．

commonly
共通にの概念が強い**一般に**を意味する副詞である．
▶ These artificial tear solutions are based on viscous polymers, most <u>commonly</u> containing methyl cellulose, hydroxypropylmethyl cellulose or polyvinyl alcohol.

comparative and superlative adjectives
一般に，**形容詞の比較級と最上級**を，er, est を付して作るか，more, most を付して作るかは，その形容詞の音節の数によって定まる．
(1)　一音節の形容詞は er, est をつける．
　　(a)　語尾が e である形容詞 rare, wide では，e を省いて考える．
　　(b)　dry は例外的に，drier, driest
(2)　二音節の形容詞は
　　(a)　語尾が y のときは，y を i に変えて，er, est をつける．
　　　　early, earlier, earliest　　　pretty, prettier, prettiest
　　(b)　その他の場合は，more, most をつける．
　　(c)　どちらの形も使用される形容詞もある．
　　　　common　　　　　　likely　　　　　　narrow
　　　　obscure　　　　　　remote　　　　　　simple
(3)　三またはそれ以上の音節からなる形容詞は
　　(a)　一般に，more, most をつける．
　　(b)　二音節の形容詞に接頭辞 un がつけられた場合は，どちらの形も使用される．
(4)　不規則な比較級と最上級をもつ例

good, better, best　　　　bad, worse, worst
far, farther, farthest, far, further, furthest

(5) 一般に二語からなる形容詞には，
　(a)　more, most を付する．
　(b)　例外は，well-known, long-lasting, short-lived など．well-known の比較級は better-known，最上級は best-known である．

comparative and superlative adverbs

(1) 一般に，**副詞の比較級**には more を，**副詞の最上級**には most をつけるが，形容詞と異なり後者に the をつけない．
▶ Certain crystalline oxides in particular, perhaps <u>most</u> notably those of transition metals, can occur in a remarkably wide range of composition.

(2) 形容詞と同じ形である次の副詞は，形容詞と同じ扱いがされる．
　close　　　deep　　　　early　　　　hard
　long　　　 low　　　　 near　　　　 wide

(3) well の比較級は better, 最上級は best（形容詞 good と同じ変化）
　badly の比較級は worse, 最上級は worst
▶ In the laboratory pure primary amines are <u>best</u> prepared by the reaction between potassium phthalimide and an alkyl halide to give an N-alkylphthalimide, which is then cleaved to give the corresponding primary amine (the Gabriel synthesis).

comparatively

厳密には比較を表し，**比較的**，**割合に**を意味する副詞である．**いくぶん**，**やや**を意味する rather, somewhat の代わりに用いることは避ける．
▶ It is a <u>comparatively</u> straightforward matter to determine standard enthalpies and Gibbs energies of formation and absolute entropies for pairs of ions in solution, for example, for a solution of sodium chloride.

compare, comparison

比較は文法上対等な関係にあるものの間に，類似性または差異を見いだす表現で，色々な形式で行われる．

(1) compare to は類似性，compare with は差異を指摘するとされるが，両者を峻別することは難しい．
▶ <u>Compared to</u> water with an effective proton affinity range of 58 kJ, the

effective proton affinity range for ammonia is more than twice as great.
- ▶ For most molecules the spacing of these allowed rotational-energy levels is quite small compared with a room temperature value of kT.

(2) as compared with は…と比較して，比較すればの意味で使用される．
- ▶ The reduced electrostatic repulsion and increased van der Waals' attraction among chalcogenide ions as compared with oxide ions are commonly given as reasons for the absence of close packing.

(3) in comparison with も (2) と同様の意味で使用される．
- ▶ The electron–nuclear magnetic interaction energy is small in comparison with the electron spin resonance energy and is independent of the strength of the magnetic field.

比較に用いられる表現には，contrast to, different from, identical to, identical with, relative to, similar to (like), unlike などがある．それぞれを見よ．

complex, complicated

ともに**複雑な**の意味の形容詞である．前者は理解には深い知識を必要とする場合に，後者は説明，分析，理解が難しい場合に用いるとされる．なお，光線の意味での light は不可算名詞である．
- ▶ The characteristic ruby-red color of Al_2O_3–Cr^{3+} in transmitted light is actually a complex mixture of blue and orange-red light.
- ▶ The mathematical treatment of conductivity is somewhat complicated, and only a very general accounts of the main ideas can be given here.

compose

…を構成するの意味の他動詞であるが，通常は**…からなる**の意味の be composed of の形で用いられ，その後に構成材料が記される．
- ▶ Many substances that appear to be amorphous are actually composed of microcrystalline units.

comprise

…からなる，含むの意味の他動詞で，be composed of, contain, consist of と同義である．
- ▶ The interstitial structures comprise the compounds of certain metallic elements, notably the transition metals and those of the lanthanide and

actinide series, with the four nonmetallic elements, hydrogen, boron, carbon and nitrogen.

上例のように能動態を用いるのが正しく，次の受動態の使用は誤りである．

▶ Alkanes are comprised of σ bonds which are unpolarized.

concentration

(1) **モル濃度**（mol L^{-1}）には，ローマン体の大文字 M を，**質量モル濃度**（mol kg^{-1}）には，イタリック体の小文字 m を用いる．

(2) 関係式中の濃度は，元素記号や化学式を角かっこ（ブラケット）［ ］に入れて表す．この記号は本文中では使用しない．

$$K_w = [H_3O^+][OH^-] = 10^{-14} \qquad v = k[CH_3CHO]^{3/2}$$

(3) 二語またはそれ以上の語からなる濃度を表す複合形容詞にはハイフンを用いる．

  one-ten-thousandth-molal solution

しかし，数値と M, m, mol dm^{-3}, または mol ％の二語からなる場合には，ハイフンを使用しない〔☞ hyphen (17)〕.

  a 0.1 M NaCl solution   a 0.1 mol dm^{-3} solution

conclusion

結論を述べるときは，現在時制，過去時制のどちらを用いてもよい〔☞ past tense, present tense〕.

consist of

部分，要素からなるの意味で，be composed of と同義である．

▶ A catalytic reaction must consist of a series of steps, each more rapid than the uncatalyzed reaction.

constant

物理定数はいずれもイタリック体またはギリシャ文字で表す〔☞ italic type, Greek letter〕.

atomic mass constant	m_u	Avogadro constant	N_A
Bohr magneton	μ_B	Boltzmann constant	k
Faraday constant	F	fine structure constant	α
gas constant	R	gravitational constant	G
permeability of vacuum	μ_0	permittivity of vacuum	ε_0
Planck constant	h	Rydberg constant	R_∞

speed of light in vacuum　c_0　　　Stefan–Boltzmann constant　σ

contrary

正反対を意味する名詞で，使用された文の他の部分と文法上，関連のない表現を形成する〔☞ comparison と比較せよ〕．

(1)　**それどころか**の意味で，on the contrary の形で用いるときには，常に文頭におく．

▶ On the contrary, single-layer graphite is accepted to have a single, characteristic HOMO level regardless of edge structure, and experiments show a single bond length.

(2)　on the other hand の意味で on the contrary を用いるときは文中におく．

contrast

対照（する），対比（する）の意味の名詞または動詞である．

(1)　**対照的に正反対なもの（人）**を意味して，to を伴う．例文中の stability は不可算名詞である．

▶ Being a model of both nuclear and electronic stability, helium provides a complete contrast to the chemical variety of hydrogen: indeed, much of its interest lies in its very inertness.

(2)　by contrast は**対照してみると**を意味する．

▶ By contrast, lithium salts of large, nonpolarizable anions such as ClO_4^- are much more soluble than those of the other alkali metals presumably because of the high energy of solvation of Li^+.

(3)　by contrast with は**…との対照によって**を意味し，比較を導入する節に用いられる．文の主語を引き合いに出すものでなければならない．

▶ By contrast with the limited studies that combine structural with thermodynamic information for discontinuous transformations, a very large body of work has been carried out on transformations in solids which are, or appear to be, continuous.

(4)　in contrast to は **…と対照をなして**の意味で，(3) の by contrast with と同じ注意を要する．

▶ In contrast to single metal centers, metal surfaces and clusters present possibilities for binding small molecules to more than one atom simultaneously.

(5) in contrast の形でも用いられる．
- ▶ <u>In contrast</u>, inorganic chemistry has a wide variety of structural types to consider, and even for a given element there are many factors to consider.

(6) 現在分詞 contrasting は形容詞として用いられる．
- ▶ The <u>contrasting</u> tendencies of benzene and borazine toward addition vs. aromatic substitution are illustrated by their reactions with bromine.

correspond to, with
相応する，一致するの意味では，to と with のどちらを用いてもよい．
相当する，該当するの意味では to を用いる．
- ▶ In general, the number of chelate donor groups that can combine with a metal ion <u>corresponds to</u> the coordination number of that ion.

counter
(1) 反，逆，対向の意味の接頭辞としては，一般にハイフンは用いない．

 counteraction   counteractive   counterbalance
 counterclockwise   countercurrent   counterflow

(2) 次の用語では，counter は形容詞として用いられている．

 counter diffusion   counter electrode
 counter electromotive force   counter ion

crystal lattice
結晶格子に用いられる略語をまとめる．

 body-centered cubic   bcc
 face-centered cubic   fcc
 cubic close-packed   ccp
 hexagonal close-packed   hcp

crystal plane and direction
(1) **Miller 指数**は一組の**面指数**で**結晶面**を表す．イタリック体の小文字を用いて，(hkl) と記し，その面に垂直方向を [hkl] と記す〔☞ bracket〕．面 (hkl) およびそれと結晶学的に同等な面をまとめて {hkl}，垂直方向をまとめて ⟨hkl⟩ で表す．なお，-1 は $\bar{1}$ と，$-h$ は $\bar{h}$ と記す．

(2) 元素と組み合わせるとき，
 (a) 元素名を用いた場合には，かっこ ( ) との間はあける．
  copper (111)   silicon (111) surface

(b) 元素記号を用いた場合には，かっこ（ ）との間を詰める．
　　Cu(111)　　　　　　　　　　Si(111) surface
(3) (hkl) 面からの **Bragg 反射**は，かっこなしで hkl と記す．例えば，(002) 面からの反射は 002 で表す．

crystallographic groups and space groups

結晶は 32 種の**点群**（**結晶群**）〔☞ symmetry elements and point group〕に，らせん操作と映進を加えて得られる 230 種の**空間群**によって分類される．これらを表す **Hermann–Mauguin の記号**（結晶群および空間群）には，イタリック体の文字とローマン体の数字を用いる．**Schönflies の記号**との対応のため，これをかっこ（ ）内に示す．

結晶群		前記結晶群から導かれる空間群の例
1	(C_1)	$P1$
$\bar{1}$	(C_i)	$P\bar{1}$
$2/m$	(C_{2h})	$P2/m$, $C2/c$
$4mm$	(C_{4v})	$P4mm$, $I4_1cd$
222	(D_2)	$P222$, $C222$, $F222$, $I2_12_12_1$
$6/mmm$	(D_{6h})	$P6/mmm$, $P6_3/mmc$
23	(T)	$P23$, $F23$, $I2_13$
$m3$	(T_h)	$Pm3$, $Ia3$
432	(O)	$P432$, $F432$, $I4_132$
$m3m$	(O_h)	$Pm3m$, $Fm3m$, $Im3m$

D

dash
en dash（二分ダッシュまたは半角ダッシュ）と em dash（全角ダッシュ）の2種類が用いられる〔☞ em dash, en dash〕.

data
datum の複数形であるが，data を不可算名詞と見なして，動詞の単数形と組み合わせる扱いも広く行われる．
▶ No data is available for comparison.
しかし，科学論文では複数形として扱うことが多い．
▶ X-ray powder data obtained from the cubic form of O_2PtF_6 were consistent with the presence of O_2^+ and PtF_6^- ions.

day
実験の記述では d と略すが，それ以外では略記しない．

deal with
取り扱う，論じる，処理するの意味である．
▶ Kinetics deals with the rate of chemical reaction, with all factors which influence the rate of reaction, and with the explanation of the rate in terms of the reaction mechanism.

despite
にもかかわらずの意味の前置詞で，in spite of と同義である．
▶ Despite both its stability and its kinetic sluggishness, dinitrogen is not totally unreactive.

differ from
二者が異なることを述べる表現である．
▶ Fluxional molecules differ from other stereochemically nonrigid molecules in possessing more than a single configuration representing an energy minimum.

different from, different to
両者は同じ意味であるが，前者の方が好ましい．そして，different と from との間に語を挿入することは避ける．したがって，次の例は望ましいものでは

ない．the dimer formed in ethanol is ... と続けるのがよい．
- ▶ The spectrum of the dimer in EtOH is red-shifted with respect to the monomer, indicating that the geometry of the dimer is <u>different</u> in ethanol <u>from</u> that formed in water.

disperse
コロイド粒子からなるの意味の形容詞であり，**分散させる**，**分散する**の意味の動詞でもある〔☞ adjective (2)，(3)〕．

disperse dye	分散染料
dispersed system	分散系
dispersing agent	分散剤

- ▶ In emulsions the droplets are referrd to as <u>disperse</u> or <u>dispersed</u> phase and the second liquid as the dispersion medium.

double negative
置き換えが可能ならば，ACS は**二重否定**よりは肯定的な表現が好ましいとする．前者の not uncommon は common で置き換えられる．
- ▶ It is <u>not uncommon</u> that a crystal face is parallel to one or two of the crystallographic axes.
- ▶ The structures of MQ_3 (M = Ti, Zr, Hf, Q = S, Se) are difficult, if <u>not impossible</u>, to describe in terms of the anion arrangement because of their low dimensionality and inherent complexity.

due to
due は**帰すべきで**，**起因して**の意味の形容詞であるから，due to の使用は be 動詞の後におく場合と，すぐ前の名詞または代名詞を修飾する場合に限るのが正しい〔☞ adjective (1)〕．
- ▶ To overcome the difficulties <u>due to</u> polarization it is useful to make use of alternating current in the determination.

したがって，due to を文頭においた次の例は正しい表現とはいえない．
- ▶ <u>Due to</u> isomorphous substitution in the clay structure of one atom with another of lower valency, the clay possesses a net negative charge.

この場合には，on account of，because of，owing to，あるいは by reason of を用いる．

E

each
(1) **おのおの，めいめい**の意味の形容詞として，一集団のすべての人またはものについて記すとき，可算名詞の単数形の前に用いる．
- ▶ The theory of membrane electrodes is quite complicated and is different in detail for each type of electrode.
- ▶ When the vibrational spectrum of a gaseous heteronuclear diatomic is analyzed under high resolution, each line is found to consist of a large number of closely spaced components.

上例のように diatomic は形容詞としてのみならず，名詞としても用いられる．次の例についても同様である．
- ▶ The large cohesive energies can account for the abundance of alkali halide clusters in alkali halide vapors: small clusters, to $n = 3$ or higher, have long been detected along with diatomics in the equilibrium vapor of the molten salt.

(2) 代名詞として用い，each of の形で，名詞や代名詞の複数形の前におく．これに続いて用いる動詞は単数形となる．
- ▶ Each of these diastereoisomers has a corresponding optical antipode.
- ▶ Each of these vibrations, called modes of vibration, corresponds to a vibrational degree of freedom.

(3) 否定の場合は，each, each of ではなく，none of を使用する．

effect
affect, effect を見よ．

e.g.
for example と読まれるラテン語 exempli gratia を略記したもので，ローマン体を用い，その前後にコンマをおく．ACS は本文中では for example を，図表の説明や本文中のかっこに囲まれた部分では e.g. を用いるとする．
- ▶ In the ionic model the valence band is made up of the top occupied anion orbitals, e.g., oxygen 2p in oxides, and chlorine 3p in chlorides, and the conduction band is composed principally of the lowest empty cation orbitals,

e.g., 3s in compounds of sodium or magnesium.
either
(1) 可算名詞の単数形の前におき，**二人（二つ）のどちらも**を意味する．
- Niobium was formerly called columbium but is probably equally unfamiliar under either name.
- Transitions from the ground state to either excited state are possible.

(2) either の代わりに，either of の形で**どちらも**の意味に用いる．この場合，of の次の名詞は複数形を，動詞は単数形を用いる．
- It is well known that the addition of hydrogen bromide to an olefinic compound may occur by either of two quite distinct types of mechanism, determined largely by the experimental conditions.

(3) **いずれか一方**の意で使われる．either A or B を主語とする場合，動詞の数はより近い位置を占める B に一致させる．
- Either the intermolecular forces of attraction and repulsion or the potential energy is dealt with as a function of the distance between the molecules.

A と B は名詞に限らないが，文法上，対等の関係にあることを要する．
- Tungstates enter into many condensed complexes, either alone or with other oxyanions, forming heteropolyacids and salts.
- Oxidation of almost all organic compounds involves either gaining oxygen atoms or losing hydrogen atoms.
- Calcium concentration, for example, is controlled either by a precipitation or by a complex-formation equilibrium to give metal ion buffering.
- The kinetics of many reactions in the solid state may well be largely dependent upon defects acting either at crystal surfaces or within crystals as centers of migration.
- The solubility of a substance may either increase or decrease with increasing temperature.
- Some solids are not easily zone-refined, either because they melt with decomposition or because they sublime too readily.

接続詞 or を二度用いた次の例の対象は二つしかないが，
- A high melting point implies either a high enthalpy of melting, or a low entropy of melting, or both.

三つのものを対象とした either A, B, or C の例も見受けられる．
▶ The M^{3+} ions form well-defined complexes—<u>either</u> anionic, cationic, <u>or</u> neutral.

electricity and magnetism
電磁気学に用いる物理量の記号と SI 単位については付表 3 を見よ．

electrochemistry
電気化学に用いる物理量の記号と SI 単位については付表 4 を見よ．

electronic shell
電子殻については atomic orbital を見よ．

electronic state
原子と分子の**電子状態**に関する記号をまとめる．
(1) 原子の電子状態の L—S 結合に基づく分類では，**全軌道角運動量** $L = 0, 1, 2, 3, ...$ に対応するローマン体の大文字の記号

   S P D F G H I K

を用い，**スピン多重度**は左上に，**全角運動量量子数** J は右下に付記する．
(2) 分子の電子状態に関しては，次の三通りの記号が用いられている．
 (a) **一重項状態**をローマン体の大文字 S，**三重項状態**を同じく大文字 T で表す．基底状態を S_0 として，励起一重項状態はエネルギーが低い方から順次，$S_1, S_2, ...$，励起三重項状態は $T_1, T_2, ...$ とする．
 (b) **非縮重状態**はローマン体の A または B，**二重縮重状態**は E，**三重縮重状態**は T で表す．スピン多重度は左上に記する．そのほか，全電子波動関数の対称性によって，右下にローマン体の添え字 1，2，g，u をつける．
 (c) 二原子分子を含め，直線分子の電子状態には，原子の電子状態に類似の記号を用いる．**全軌道角運動量** $\Lambda = 1, 2, 3, 4$ に対応する大文字のギリシャ文字は，Σ，Π，Δ，Φ である．スピン多重度は左上につけ，対称中心をもつ分子の Σ 状態は，分子軸を含む平面での鏡映によって，$+$，$-$ に分け，これらを右上につける．電子状態間の**遷移**を記載する場合は，**吸収**と**発光**を矢印の向きで区別する．
▶ The absorption spectrum of H_2 shows bands corresponding to the $^1\Sigma_u^+ \leftarrow\ ^1\Sigma_g^+$ and $^1\Pi_g \leftarrow\ ^1\Sigma_g^+$ transitions. In a hydrogen arc lamp, molecules are produced in the $^3\Sigma_g^+$ state, and the characteristic emission is due to the $^3\Sigma_g^+ \rightarrow\ ^3\Sigma_u^+$.

element

(1) **元素名**はローマ体で記し，普通名詞として扱う．

(2) **元素記号**はローマ体で記し，頭字を大文字とする．

(3) 元素記号が元素を表す場合には，その元素名として読む．例えば，Au は gold，N は nitrogen と読む〔☞ a, an (6)〕．

(4) 核化学の考察では，原子番号を元素記号の左下に付記する．

(5) **同位体**を指定するには，質量数を元素記号の左上に付記するか，元素名の後に記載して，両者をハイフンで結ぶ．これらを読むには，元素名または元素記号からはじめる〔☞ a, an (6)，chemical equation (5)〕．

^{35}Cl   chlorine-35

(6) 元素記号と反応の型を表す名詞，または形容詞が組み合わされたときには，ローマ体の元素記号と次の語の間にハイフンを入れる．

N-alkylation   O-substitution   S-methylated

em dash

全角ダッシュの前後は詰める．

(1) 挿入句や節の前後に，全角ダッシュが用いられるが，コンマを用いる方がより望ましい〔☞ comma (3)〕．

▶ Werner's coordination theory had an inherent weakness in that it postulated two different kinds of valence for inorganic substances—primary and secondary valence bonds—without any theoretical justification for their existence.

(2) ハイフンを用いた雑誌名の略記に用いられる．

Chem.—Eur. J.   *Catal. Rev.—Sci. Eng.*

J.—Am. Water Works Assoc. *J.—Assoc. Off. Anal. Chem.*

en dash

二分ダッシュの前後は詰める（演算子－の場合は前後をあける）．

(1) 対等な関係にある二語からなる概念において，and，to，versus の意味で用いられる．

alkyl–heavy metal   calcite–aragonite transformation
d–d transition   dipole–induced-dipole interaction
donor–acceptor complex   high-potential iron–sulfur proteins
gas–solid reaction   ion–dipole interaction

keto–enol equilibrium　　　　　　nitrito–nitro isomerization
sol–gel process　　　　　　　　　spin–lattice relaxation time

ただし，色の組合せにはハイフンを用いる〔☞ hyphen (7)〕．

(2)　三個以上の数や範囲を表す場合には

　(a)　to, through の意で二分ダッシュを用い，数字との間は詰める．上付きの引用文献番号についても同じ．

　1–3　　　　　　　　　　　　　　10–100 ℃
　10–20 mg　　　　　　　　　　　Figures 1–3

　(b)　マイナスなどの記号を伴うときには，二分ダッシュではなく to や through を使用する．

　−10 to ＋100℃　　　　　　　　−100 to −30 ℃
　≈40 to 50　　　　　　　　　　　10 to＞100 mL
　＜10 to 20 mg

　(c)　from ...to ..., between ...and ... における to, and の代わりに二分ダッシュを用いることはしない．

　　　from 10 to 100 mL　　　　　（from 10–100 mL は不可）
　　　between −10 and ＋100 ℃　（between −10–＋100 ℃は不可）

(3)　対等な関係にある複数の人名を形容詞として用いるときには，二分ダッシュを用いる．これらの人名を冠した方法，反応などには，定冠詞 the をつけるのが通常である．

　Debye–Scherrer method　　　　Diels–Alder reaction
　Joule–Thomson effect　　　　　Russell–Saunders coupling

(4)　混合溶媒の成分の間をつなぐに用いる．

　ethanol–water　　　　　　　　　hexane–benzene

この場合は，二分ダッシュの代わりに，スラッシュを用いてもよい．

　ethanol/water　　　　　　　　　hexane/benzene

(5)　混合物，固溶体など，多成分系の成分の間をつなぐのに用いる．

　CaF_2–YF_3　　　　Ca–Si–O　　　　Al–Cu–Zn

担持触媒では，担体の前にスラッシュが用いられるから，二分ダッシュとスラッシュは併用される．いずれにおいても，Al_2O_3 は担体である．

　Cr_2O_3/Al_2O_3　　　　　　　K_2O–Cr_2O_3/Al_2O_3

equal

(1) 本文中，等号＝を is や equals の略号として用いてはいけない．

(2) equal か unequal かであって，これらの形容詞に中間の程度はない．厳密に等しくはないときには，approximately equal, more nearly equal, less than equal などの表現を用いる．ただし，次の例も見受けられる．

▶ With the correct level of electrolyte added to a solution of an anionic surfactant, cylindrical micelles can be induced to grow to lengths <u>equal to</u> many times their diameter.

同様の扱いを受ける形容詞には，次のものがある．

circular, even, perfect, unique, round, square

equation

数式に使用する書体に関しては，italic type, roman type を，記号に関しては，symbols を参照せよ．

(1) 一度しか使用されない短い数式は，本文の行中に記載してよい．

(2) 文の一部として数学的表現を用いる場合は，主語，動詞，目的語すべてが数学的表現であることを要する．下例では，主語は V，動詞は＝，目的語は 12 である．

When $V = 12$

(3) 改行して記載された数式によって，節や文が完結する場合には，数式の前にコロンを使用しないように注意せよ．

(4) 改行して記載された数式の後には，終止符をつけない．

(5) 繰り返し言及される数式は，改行して記載し，順番に番号をつける．

(6) equation が文頭にあって番号を伴わないときには，これを綴る．本文中では eq（複数形は eqs）を用い，eq 2, eqs 3 and 4 のように記載する．

(7) 演算子＝，＞，＋，－，÷などの前後をあける．演算子－と二分ダッシュとは，これによって区別される．

$x = a$ (x is equal to a, or x equals a)
$x > a$ (x is greater than a)
$b < 5$ (b is less than five)
$a \propto b$ (a is proportional to b)

ただし，形容詞の働きをしている記号と数字の間は詰める．

$+10$ -20 <30 ≈ 40

(8) 三角関数，対数，指数などローマン体の関数の前後はあける．ただし，これらにかっこ（ ）が続く場合は，その間を詰める．

$\sin x$ $\qquad\sin(x+1)$
$\log x$ $\qquad\log(x/a)$
$\exp(-x)$ $\qquad\mathrm{pH}=-\log a(\mathrm{H}^+)$

(9) 微分，dx，dy の前後はあける．

(10) 数字と変数，変数と変数の間は詰める．

$2x$ $\qquad xy$
$a_n a_{n+1} a_{n+2}, ...$ $\qquad PV=nRT$

(11) /，:，・，( ) の前後は詰める．スラッシュを用いた場合，必要ならば，かっこ（ ）を用いて，分母子の区別が明確になるように注意せよ．

a/b $\qquad a:b$
$a \cdot b$ $\qquad 3(xy)$
$(3x)y$ $\qquad (a/b)/c$
$(a/b)(c/d)$ $\qquad (x-y)/(2x+y)$
$(x+y)/2=z$ $\qquad [(x+y)/2]+2z$

(12) 記号と上付き文字，下付き文字の間は詰める．上付き文字，下付き文字それぞれも，混乱の恐れがない限り詰める．

etc.

…など，**その他**を意味する and the rest, and others, and so forth に相当するラテン語 et cetera を略記したもので，ローマン体を用いる．コンマの使用法は and の場合と同様で，A etc. や A, B, etc. とする．これらの例から明らかなように，etc. の前に不用なコンマを入れないよう注意する．

▶ A large class of aromatic nitro compounds, quinones, carboxylic acids, sulfonic acids, etc., will combine with aromatic hydrocarbons, amines, phenols, and related substances to give stable solid complexes which are frequently intensely colored.

ACS は etc. を図表の説明や本文中のかっこで囲まれた部分でのみ使用し，本文中では and so forth を用いるとする．

every

可算名詞の単数形の前において，**一集団の個々全部**を意味する．every の代わりに all を用いるときには名詞は複数形にする．個々の意味では each を用

いる.
- ▶ The diamond lattice is constructed to allow every atom to complete its electron octet.

except
(1) **…のほかは**の意味の前置詞で，but よりも強い除外を意味する．
- ▶ We usually consider structures of MX-type solids in terms of packing of ions, except where covalent bonding is obviously of primary importance.

(2) except for は**…がなければ，…があるだけで**の意味である．
- ▶ In general the vibrational-entropy contribution is small but, except for wide vibrational spacings as in CO, not negligible.

excited electronic state
励起電子状態を表すには化学式の右にアステリスク＊をつける．
$$He^* \qquad NO^*$$

exposing, be exposed, on exposure
光にさらす，感光させるなどの表現に用いる．
- ▶ Raman spectra are produced by exposing a gas, liquid, or solid to bright light and examining the spectrum of the light scattered laterally.
- ▶ When a photographic film is briefly exposed to light, some of the grains of silver bromide undergo a small amount of decomposition.
- ▶ When, on exposure of the emulsion to light, a photon of energy impinges on a grain of AgX, a halide ion is excited and loses its electron to the conduction band, through which it passes rapidly to the surface of the grain where it is able to liberate an atom of silver.

extreme, extremely
非常なの意味の形容詞と，**非常に**の意味の副詞である．
- ▶ Apart from their very high melting point and extreme hardness, all of these interstitial carbides, nitrides, and borides are extremely inert chemically.

F

fall on
(光が) …に当たるの表現に用いられる．
- Hertz found that when suitable radiation <u>falls on</u> a metal surface, electrons are emitted from the surface.

fewer
数の比較に用い，名詞の複数形と組み合わせる〔☞ less と比較せよ〕．

figure, number
前者は**数字**を指し，後者は**数**を指す〔☞ number, numeral〕．
- Introduction of potassium between graphite sheets increases the normal interplane distance of 3.35 to 5.40Å, the corresponding <u>figures</u> for rubidium and cesium being 5.61 and 5.95Å, respectively.

fill
満たす，一杯にするの意味の他動詞である．特に受動態を用いるときには，満すもの，満されるものがそれぞれ何であるかに注意せよ．
- Cations <u>fill</u> the interstices between adjacent anion layers, which means that in these hcp or ccp packing schemes the coordination geometry of a centered cation is limited to octahedral or tetrahedral.
- The pycnometer <u>is filled</u> slightly beyond the file mark with the liquid the density of which is to be determined.
- The most common form of TiO_2, rutile, consists of a distorted hcp array of oxide ions with half of the octahedral interstices <u>filled</u> in an ordered way by Ti^{4+} ions.

find, discover
見つけるの意味では前者が一般的である．発見の意味では後者を用いる．
- The cathode rays <u>were discovered</u> by J. Plücker, professor of physics in Bonn, who <u>found</u> that they were deflected by a magnetic field.

first
第一，最初の意味の形容詞，副詞，代名詞，名詞として用いられる．
- The rules should evolve as new structures become available, but we have

found that the most important relationships are usually seen from the first 10 or so structures in a series.

▶ The most stable hydride of P is phosphine (phosphane), PH_3. It is the first of a homologous series P_nH_{n+2} ($n = 1–6$) the members of which rapidly diminish in thermal stability, though P_2H_4 and P_3H_5 have been isolated pure.

flammable, inflammable

いずれも**可燃性の**, **燃えやすい**の意味の形容詞である．**不燃性の**の意味の形容詞は nonflammable である．なお，heat は物質名詞であるが，heat of combustion は普通名詞として扱われる．

▶ B_2H_6 is spontaneously flammable; it has a higher heat of combustion per unit weight of fuel than any other substance except H_2, BeH_2, and $Be(BH_4)_2$.

fold

(1) **倍**の意味の接尾辞で，一般にハイフンは使用しない．

    twofold   sixfold   manifold

(2) 10 倍以上の場合は数字を用いて，ハイフンを使用する．

    10-fold   20-fold

follow

(1) **…に続く**, **…の結果として起こる**の意味の動詞である．

▶ From the relationship between ΔG and the cell emf, it follows that the more positive the oxidation potential of a half reaction, the greater will be its thermodynamic tendency to take place.

▶ Proteins are structurally best characterized by X-ray analysis following crystallization.

▶ All preparations containing silver were analyzed by treatment with boiling aqua regia, followed by filtration, washing, and weighing of the resulting AgCl.

(2) as follows の動詞は，名詞の複数形や二つ以上の名前，もの，語が関与する場合も，as follow と変化しない．

▶ The compositions of the products were checked by chemical analysis as follows: ....

(3) following は**…について**, **…の後での**意味の前置詞としても用いる．

▶ Following treatment in O_2, the Pt particles on TiO_2 were large and dense, i.e., similar to Pt on SiO$_2$, while treatment in H$_2$ reduced TiO$_2$ to Ti$_4$O$_7$ and caused the metal particles to spread into thin, flat structures.

former, the, latter, the

対句 the former ..., the latter として用いるとは限らない．二人，二つのもの，二つの集団などについて述べるときに用いる．

▶ In the latter, a diradical is formed initially, followed by a hydrogen atom transfer in a separate stage.

三つ以上について述べるときには，first と last を用いるのがよい．

fraction

(1) **分数**は，分母子ともに 10 以下の場合は，略記しないで綴り，ハイフンを使用する〔☞ hyphen (15)〕．

  one-half   one-quarter
  two-thirds   three-fourths

(2) 分母子のいずれかが 10 以上の場合は，数字を用い 1/10，13/20 などと表現する．スラッシュの前後は詰める．

(3) 分数が主語である場合，その数の扱いは

 (a) 分数の次の of に続く語が単数か，複数かによって定まる．

 (b) remainder of ..., rest of ... についても，これに準じる．

(4) 上付きおよび下付き文字では，スラッシュを用いて，1/2，2/3，a/b などとする．

  $x^{1/2}$ $(a-b)^{1/3}$ $t_{1/2}$

function

関数にはローマン体とイタリック体のものがある．

(1) 三角関数，対数，指数にはローマン体を用い，変数との間はあける．ただし，変数がかっこ（ ）の中に入っているときには詰める．

  $2\sin x$ $-\log y$ $\exp(-z)$

(2) 一般的な関数にはイタリック体を用い，かっこ（ ）の前は詰める．

  $f(x)$ $F(r)$ $V(R)$

G

gamma ray (γ ray)
　γ 線の表記は γ ray とし，gamma ray とはしない．X-ray の場合とは異なり，表題や文頭では，ray の頭字を大文字にする．γ ray を形容詞として用いたときには，γ-ray となる〔☞ Greek letter (5)〕．

general, in
　一般に，一般のを意味の慣用的表現で，generally の代わりに用いられる．同じ形式の慣用的表現に in particular，in short がある．
- ▶ In general, a substituent that stabilizes a cation more than a base will increase the ease of ionization, and hence raise pK_a, whereas if the neutral molecule is preferentially stabilized, the substituent is base-weakening.
- ▶ Atomic radii may then be subdivided into either metallic, as in metals, alloys, or intermetallic compounds, or covalent, as in nonmetals and in covalent molecules in general.

general chemistry
　一般化学に用いる物理量の記号と SI 単位については付表 5 を見よ．

generally
　一般にを意味する副詞であるが，口語では多くの場合を意味する usually の代わりに用いられるので，意味が不明確になるのを避けるには，in general を使用する．副詞は強調したい位置におく．
- ▶ It is generally recognized that vertical similarity of the elements in the periodic table is the rule with the exception of the rare earths and elements of Group VIII, for which horizontal similarity seems to apply.
- ▶ When hydrogen bonding is possible in molecular crystals, the structure is generally one that yields the maximum number of hydrogen bonds.

get, obtain
　両者ともに得るを意味するが，前者は努力，意志の有無に関係はない．後者は目的の達成を暗示する．
- ▶ It is very difficult to get the reagents to react fast enough to make the process feasible.

▶ Much more conclusive evidence for the presence of traces of enol in equilibrium with simple carbonyl compounds has been obtained from kinetic studies of the acid-catalyzed halogenations, racemizations, and deuterium exchanges of ketones.

give rise to
…を起こす，…のもとであるの意味である．

▶ Loss of an electron from the molecule followed by various fission processes gives rise to ions and neutral fragments.

▶ One observes, in fact, that molecules like H_2, N_2, and CO_2, which are linear, give rise to no absorptions that can be attributed to changes in the rotational energy of the molecules.

-gram
記録，記録図の意味の接尾辞で，名詞を作る．

| chromatogram | diagram |
| thermogram | voltammogram |

-graph
記録のための機器の意味の接尾辞で，名詞を作る．

| chromatograph | electrograph |
| polarograph | spectrograph |

-graphy
記録法の意味の接尾辞で，不可算名詞を作る．

| chromatography | crystallography |
| photography | polarography |

Greek letter
ギリシャ文字は化学用語，物理用語でしばしば使用される．

(1) 物理量を表す記号に用いられる．

α angle of optical rotation, plane angle, polarizability
β plane angle
Γ surface concentration
γ activity coefficient, mass concentration, surface tension
δ chemical shift (δ scale), distance
ε molar absorption coefficient, permittivity

ζ	electrokinetic potential
η	overpotential, viscosity
Θ	characteristic temperature, quadrupole moment
θ	Celsius temperature, plane angle
κ	(electric) conductivity, thermal conductivity
Λ	molar conductivity (of an electrolyte)
λ	decay constant, molar conductivity (of an ion), wavelength
μ	chemical potential, permeability, reduced mass
ν	frequency, stoichiometric coefficient
$\tilde{\nu}$	wavenumber in vacuum
ξ	extent of reaction
Π	osmotic pressure
ρ	charge density, mass concentration, resistivity
σ	surface tension, symmetry number
τ	time constant, transmittance
Φ	magnetic flux, work function
ϕ	plane angle, quantum yield, volume fraction
χ	magnetic susceptibility
Ω	electric resistance, solid angle
ω	circular frequency, solid angle

(2) 核子の種類を表すに用いる〔☞ alpha particle, beta particle〕.

　　α particle　　　　　　　　β particle

(3) 結合軌道およびそれによって生じる結合に用いる〔☞ bond (2)〕.

　　σ orbital　　　　　　　　π bond

(4) 化学物質名において，位置を示すのに使用される．

　　α-hydroxy-β-aminobutyric acid

　　β-chloro-1-naphthalenebutanol

(5) ギリシャ文字を冠した名詞を形容詞として使用するときには，ハイフンを必要とする〔☞ hyphen (5)〕.

　　γ-ray spectrometer　　　　π-electron system

H

hardly
否定的意味をもつ副詞であるから，肯定的な語と組み合わせる．
▶ The situation is hardly clear-cut; we can write the hydrated proton as H^+, H_3O^+, $H_9O_4^+$, or $H_{11}O_5^+$.

hence
それゆえにを意味する副詞で，consequently, therefore, so that よりは格式ばった表現とされる．
▶ Molybdic acid, H_2MoO_4, cannot be formed from MoO_3, although the reverse is possible; hence, molybdenum trioxide is not strictly an anhydride.

hetero
(1)　**異種の**の意味の接頭辞で，一般にハイフンは用いない．
　　heteroatom　　　　　　　　heterocyclic ...
　　heterogeneous ...　　　　　heteronuclear ...
(2)　形容詞として扱われることもある．
　　hetero group

highest occupied molecular orbital
最高被占軌道は HOMO と略記する．すべて大文字からなる略号であるから，複数形は小文字の s だけをつけた HOMOs である〔☞ abbreviation (12)〕．

homo
(1)　**同一の**の意味の接頭辞で，一般にハイフンは用いない．
　　homogenous　　　　homologue　　　　　　homolysis
(2)　形容詞として扱われることもある．
　　homo nucleoside

however
しかしながらの意味の副詞である．
(1)　導入の語として文頭で使用した場合には，次にコンマを要する〔☞ comma (2)〕．
▶ However, the choice of the metal used for the electrodes can also affect the various electrode processes, and this increases the separation factor,

$(H/D)_g/(H/D)_l$, still further.

(2) 二つの節の間におくときは，前はセミコロン，後はコンマで区切る．

▶ X-ray crystallography is clearly the most powerful method for determining the conformations of the molecules; <u>however</u>, again the conformation of the molecules in the crystal may not correspond to the biologically relevant conformations in solution or in the biological milieu.

(3) その他の位置では，前後ともハイフンで区切るのが正しいとされる．

▶ We shall see, <u>however</u>, that Hess's law of heat summation is merely an application of the first law of thermodynamics.

hence, moreover, nevertheless, nonetheless, therefore, thus も同じ取り扱いを受ける．ただし，必ずしも実行されていない．

hyphen

行末で語が分割されるときの**ハイフン**の用法については syllabication を，接頭辞と接尾辞における用法については prefix (2)，suffix を見よ．

(1) **複合語**にはハイフンを用いるものと，background, crosslinkage のように，一語で書くものとがあるので辞典を参照せよ．

after-treatment	cross-propagation
glass-ceramic	half-life
network-former	off-flavor
O-ring	rust-proofing
self-consistent	X-ray

(2) 元素記号と名詞または形容詞を組み合わせるときは，ローマン体の元素記号と次の語の間にハイフンを入れる〔☞ element (6)〕．

N-alkylation	S-bonded
^{14}C-labeled	N-oxidized

(3) 二語からなる動詞にはハイフンを用いる．

air-dry	blue-shift（名詞は blue shift）
freeze-dry	ring-expand

他方，動詞と副詞が結合した**句動詞**の場合にはハイフンを用いない．

break down	fall off
mix up	set off

しかし，これらも名詞や形容詞として用いるときには，ハイフンでつないだ

り，間を詰めて一語とする．

    breakdown   falloff

    fall-off curves   mix-up

（4）　姓名におけるハイフンは本人の使用に従う．

    Joseph-Louis Gay-Lussac   Irène Joliot-Curie

    Jean-Marie Lehn   John Edward Lennard-Jones

（5）　二語がまとまった**複合形容詞**には，語間にハイフンを入れる．それには，色々な組合せがある．例外は hyphen（6）を見よ．

名詞と形容詞，名詞と分詞からなる句は，be 動詞の後にも使用される．

  （a）　名詞と名詞の組合せ

    argon-ion laser   bond-moment analysis

    charge-transfer complex   collision-theory approach

    electron-pair acceptor   electron-transfer reaction

    energy-level pattern   energy-transfer mechanism

    gas-phase reactions   hydrogen-atom abstraction

    hydrogen-bond pattern   hydrogen-ion concentration

    ion-combination reaction   ion-exchange resin

    ion-pair formation   ligand-field theory

    potential-energy surface   proton-switch mechanism

    room-temperature conductivity

    transition-metal complexes   transition-state theory

    vapor-phase reaction   valence-electron delocalization

  （b）　名詞と形容詞の組合せ

    carbon-rich materials   electron-deficient compounds

    energy-dispersive system   heat-sensitive device

    ion-reversible electrodes   ion-selective electrodes

    radiation-chemical processes   surface-active molecules

    time-dependent stoichiometry

  （c）　名詞と**現在分詞**（他動詞の目的語が現在分詞の前におかれる例が多い）の組合せ（ただし，現在分詞とそれに続く名詞の組合せに意義がある場合には，形式は似ていてもハイフンを用いない．例えば，iron printing process，ligand binding site）

chain-propagating reaction
complex-forming supporting electrolyte
flame-coating enamel
gas-washing bottle
light-emitting diodes
oil-hardening steel
temperature-increasing curve

electron-withdrawing group
fluorine-containing compounds
hydrogen-bonding liquids
milk-coagulating enzyme
rate-determining step
time-consuming process

(d) 名詞と**過去分詞**の組合せ

acid-catalyzed reactions
body-centered cubic structure
computer-assisted design
diffusion-controlled reactions
hand-ground mixtures
laser-based resonance-ionization spectroscopy
lead-glazed pottery
nickel-filtered Cu Kα radiation
singlet-excited state
spin-forbidden transitions
surface-enhanced Raman spectroscopy
temperature-programmed desorption
time-resolved spectroscopy

(e) 過去分詞または形容詞と名詞の組合せ

branched-chain reaction
condensed-phase phenomena
double-sphere model
fast-flow technique
free-radical substitution
high-energy photon
ideal-gas equation
inner-sphere mechanism
long-range order
low-resolution instrument

closed-shell atomic ions
critical-state constants
fast-exchange limit
first-order reflection
half-wave potential
high-temperature superconductors
ionic-strength method
kinetic-isotope effects
low-energy electron
low-spin electron configurations

　　　　many-electron atoms　　　　　many-body problem
　　　　minimum-energy path　　　　　noble-gas compounds
　　　　normal-mode analysis　　　　　polar-bond mechanism
　　　　right-hand subscript　　　　　round-bottom flask
　　　　second-order reactions　　　　short-range forces
　　　　small-angle X-ray scattering　 solid-state chemistry
　　　　steady-state hypothesis　　　　stirred-flow reactor
　　　　stopped-flow technique　　　　thin-layer chromatography
　　　　two-parameter equation　　　　zero-field splitting
　　(f)　形容詞と形容詞の組合せ
　　　　vibrational-rotational spectra
　　(g)　副詞と形容詞の組合せ
　　　　above-average results　　　　 direct-reading balance
　　　　far-reaching influence　　　　long-lived polymerization
　　　　less-symmetrical coordination geometry
　　　　much-investigated reaction　　 well-defined complexes
ただし，-ly でおわる副詞にはハイフンは用いない．
　　　　accurately measured values　　circularly polarized light
　　　　newly developed method　　　　optically active substance
　　(h)　数詞を含む場合
　　　　five-membered ring　　　　　　one-dimensional solids
　　　　six-coordinated structure　　　many-membered ring
この場合の二語目が名詞であっても，それを複数形にはしない
　　　　three-parameter equation　　　three-neck flask
　　　　two-component system　　　　　two-photon process
　　(i)　ギリシャ文字を含む名詞
　　　　γ-ray detector　　　　　　　π-electron system
(6)　二語がまとまって名詞を修飾する場合でも，次の場合にはハイフンを使用しない．
　　(a)　外来語〔☞ Latin term〕
　　　　ab initio methods　　　　　　 in situ evaluation
　　　　in vivo nitrogen fixation

(b) 固有名詞を含む場合
Lewis acid catalysis   Fourier transform infrared
(c) 二語からなる化学物質名
barium sulfate precipitate   potassium hydroxide solution

(7) 色を表す語の組合せ（対等な関係でも，二分ダッシュは用いない）
blue-green   bluish-purple   silver-gray

(8) ever, ill, little, still, well を含む二語からなる形容詞には，ハイフンを使用する．
well-studied hypothesis

ただし，これが副詞，very, most などでさらに修飾されるときは，ハイフンを用いない．
very well studied hypothesis

(9) 比較級，最上級を含む二語からなる形容詞
best-known process   lowest-frequency wavelength
least-squares analysis（squares が複数形であることに注意せよ）
nearest-neighbor interaction

(10) 三語以上からなる形容詞にはハイフンを使用する．既に二語の間にハイフンが使用されている場合を含む．

copper-to-iron ratio
edge-to-edge interaction
head-to-head placement
head-to-tail configuration
metal-to-ligand charge transfer
one-and-a-half bond
one-and-two-thirds bond
one-to-one relationship
particle-in-a-box problem
push-and-pull mechanism
self-consistent-field calculation
state-to-state reaction dynamics
stick-and-ball models
term-by-term calculation

three-center-two-electron bond

trial-and-error procedure

transition-metal-promoted reaction

(11) 既にハイフンを用いてある複合形容詞に，さらに接頭辞をつけるときには，ハイフンを使用する〔☞ prefix (5)〕．

pre-steady-state condition　　　pseudo-first-order reaction

(12) 既にハイフンを用いてある複合形容詞に，さらに固有名詞をつけるときには，ハイフンを除く．

acid-catalyzed reactions　　　Lewis acid catalyzed reactions

(13) 二分ダッシュで結ばれた語を含む場合

alkyl–heavy-metal complex　　high-spin–low-sin transition

metal–metal-bonded complex　Mo–Fe-containing protein

single-crystal–single-crystal transformation

(14) 複合形容詞と名詞の間にかっこ（ ）に入れた記述がある場合

a 7-in. (17.8-cm) funnel

high-temperature (200 °C) reaction

element (silicon or tin)-centered radicals

(15) 分母子ともに 10 以下の分数は，略記しないで綴り，ハイフンを使用する〔☞ fraction を見よ〕．

one-half　　　　　　　　　one-quarter

two-thirds　　　　　　　　three-fourths

(16) 倍の意の接尾辞 fold の使用において，10 倍以上の場合は数字を用い，fold との間にハイフンを使用する．10 倍以下の場合は綴る〔☞ fold〕．

10-fold　　　　　20-fold　　　　　ninefold

(17) 数と単位からなる形容詞にはハイフンを使用する．二つの数の間に二分ダッシュまたは to が用いられているときも同様である．

10-min exposure　　20-mg sample　　a 20–25-mL aliquot

a 1- to 2-h sampling time　　　　a 20- to 25-mL aliquot

ただし，次の例外がある．

(a) 単位が M, m, mol dm^{-3}, mol % のときにはハイフンを用いない．

(b) 温度の単位，°C, K にはハイフンを用いない．

(c) 単位が °, ′, ″, % のときには，数との間を詰めるのでハイフンを用

いない．

(d) ×10^n をつけた数値にはハイフンを用いない．

特定の物質や溶液であれば of を用いて，次例のように表現するから，ハイフンとは無縁である．

▶ Phenyl propionate may be prepared by slowly adding 196 g of redistilled thionyl chloride to a mixture of 150 g of pure phenol and 132 g of propionic acid, warming to drive off all the sulfur dioxide and hydrogen chloride, and distilling; 190 g (79%) of phenyl propionate, bp 202 − 212°C, is obtained.

(18) 数と単位に，さらに一語が加わる場合には，二つの語間ともにハイフンを入れる．

  5-mm-thick layer   120-nm-diameter droplets

(19) 複数の位置記号や接頭辞が，対等の関係で後続する語に帰属するときは，ハイフンを使用して簡略化できる〔☞ prefix (9)〕．

  o- and *p*-nitrophenols

  pyro- and piezoelectric

  mono-, di-, and trichloroacetic acids

ただし，次の例のように，それぞれに化合物番号を記載した場合には，acid を複数形にはしない．

▶ Acetic acid can be chlorinated by gaseous chlorine in the presence of red phosphorus to yield successively mono- (**1**), di- (**2**), and trichloroacetic acid (**3**); the reaction proceeds better in bright sunlight.

(20) ハイフンを用いた複数の複合形容詞が，対等の関係で後続する名詞に帰属するときには，次のように簡略化できる．

  four- and five-membered rings

  ground- and excited-state reactivities

  X- and *γ*-ray-induced polymerization

I

identical
正確に詳細にわたって同じの意味の形容詞で，比較級はない．
- ▶ The length of a given bond is found to be almost identical in different compounds, provided that the bond is not incorporated into a conjugated system.
- ▶ Synthetic rubbers are, rather, substitutes for rubber—materials with properties and structure similar to but not identical with those of natural rubber.

i.e.
that is の意味のラテン語 id est を略記したもので，ローマン体を使用する．二つの節の間では，前にセミコロン，後にコンマを用い，それ以外の場合は，前後にコンマを用いることは，that is の扱いと変わらない〔☞ that is を参照せよ〕．ACS は図表の説明や本文中のかっこで囲まれた部分でのみ i.e. を使用し，本文中では that is を用いるとする．
- ▶ The attraction a molecule exerts on its neighbors tends to draw them in toward itself; i.e., the attraction acts with the confining pressure to hold the molecules together.
- ▶ Mössbauer spectroscopy is at its best in situations where other techniques are difficult or impossible, i.e., in the study of compounds which are insoluble solid powders.
- ▶ Bonds or groups within a molecule sometimes vibrate with a frequency, i.e., have an energy-level pattern with a spacing, that is little affected by the rest of the molecule.

if ... were
事実に反することを述べるときには過去時制を用い，主語が単数形でも動詞は was ではなく were を使用する．その場合，主節には助動詞 should, would, could, might を使用する．
- ▶ If the configuration of the Cu^{2+} ion were $d^1_{z^2}d^2_{x^2-y^2}$, we should expect an exactly opposite distortion.

importance, of
前置詞と抽象名詞からなる句で，**重要である**の意味である．
▶ Redox reactions are of wide importance in chemistry.
同じ形式の句には，つぎのものもある．
    of difficulty   of interest   of significance

in ...ing
…することにおいての意味を表す．
▶ In considering the reactive process between chemical species, several factors must be taken into account.
▶ In dealing with the reaction under conditions where the steps are second-order, there are various limiting cases that are easily solved.
▶ In determining relative strengths of the stronger acids such as HCl or H_2SO_4 in an aqueous system, the equilibrium constants are all found to be indeterminately large.

including
…**を含めて**の意味で，制限的用法の場合はコンマで区切らず，補足的説明を加える非制限的用法の場合はコンマで区切る．
▶ All molecules, including those without a permanent dipole, attract each other.

inter
（1）…**の間**または**相互関係**の意味の接頭辞で，一般にハイフンは用いない．

    interatomic   interfacial   intermolecular
    interpolation   interstitial   intersystem
    interaction   interchange

（2）rが重なる場合はハイフンを用いる．

    inter-ring

interest, of
of と interest の間に，さらに形容詞を挿入してもよい．
▶ The magnitudes of the individual ion conductivities are of considerable interest.
▶ Of particular interest is the very high conductivity of the hydrogen ion.

intra
(1) **内に，内部に，内側に**の意味の接頭辞で，一般にハイフンは用いない．

    intramolecular   intranuclear

(2) aが重なる場合は例外的にハイフンを用いる．

    intra-atomic   intra-ring

involve
伴う，関連させる，関係する，包含するの意味の他動詞である．

▶ The electrons in the valence band are not free to move because they are involved in chemical bonding.

▶ The isotherm for two substances adsorbed on the same surface is of importance in connection with inhibition and with the kinetics of surface reactions involving two reactants.

ionic charge
イオン電荷は電荷数，次いで正負の記号を書く．化学式の右上に電荷数をつけるとき，電荷数が1の場合は＋，－のみを書く．原子数を表す下付き文字があれば，これを先に書く．複数個の＋，－を用いることはしない．

    O_2^- dioxide(1−) ion O_2^{2-} dioxide(2−) ion

    $[AuCl_4]^-$ tetrachloroaurate(1−) ion

irradiate
…を照らす，…に放射線を照射するの表現に用いる．受動態を用いるときには，照射されるものが主語となる．

▶ If a mixture of dry hydrogen and chlorine is irradiated by visible light, the reaction

$$H_2 + Cl_2 \rightarrow 2HCl$$

occurs with explosive violence. The reaction between hydrogen and bromine can also be brought about by irradiating the mixture with light of suitable wavelength.

illuminate もまったく同様に用いられる．

▶ When a mixture of hydrogen and chlorine is illuminated with blue light, reaction immediately begins.

IR spectroscopy

赤外線分光法のことで，infrared の略記である IR を定義する必要はない．測定結果の記載方法は幾通りもある．

　　IR (KBr)　$\tilde{\nu}_{max}$(cm^{-1}) : 3017, 2953 (s, OH), 2855 (s), 2195, 1512, 1360, 1082, 887.

　　FTIR (CCl$_4$)　$\tilde{\nu}_{max}$: 1686 cm^{-1} (C=O).

　　IR (cm^{-1}) : 2253, 1705, 1620.

$\tilde{\nu}_{max}$ は吸収極大の波数．吸収線の形状に関する略号は br, broad ; m, medium ; s, strong ; vs, very strong ; vw, very weak ; w, weak.

　なお，near-infrared は NIR と略記されるが，IR と異なり定義の必要がある．

italic type

　(1)　広く使用されているラテン語とその略語は**イタリック体**にはしないで，ローマン体を用いる〔☞ Latin term〕．

　(2)　**物理量の記号，変数**はイタリック体とする〔☞ physical quantities, Greek letter〕．

　　activity　　a　　　　　　　heat capacity　　C
　　relative density　d　　　　electric current　I
　　equilibrium constant　K　　length　　l
　　mass　　m　　　　　　　　amount of substance　n
　　pressure　p, P　　　　　　rate of reaction　v
　　time　　t　　　　　　　　thermodynamic temperature　T
　　volume　　V　　　　　　　mole fraction　　x, y

　これらは，形容詞の一部となっても，下付き文字となっても，イタリック体であることに変わりはない．

　　i th　　　　　　　　n th　　　　　　　$C_{V,m}$
　　K_c　　　　　　　　P_i　　　　　　　　ΔS_m
　　Al$_2$(SO$_4$)・nH$_2$O　　Li$_x$NO$_2$　　　　Li(Ni$_{1-y}$Co$_y$)O$_2$

　(3)　**定数**にはイタリック体またはギリシャ文字を用いる〔☞ constant〕．人名に由来する下つき文字は，ローマン体とする．

　　Avogadro constant　N_A　　　Bohr radius　a_0
　　Boltzmann constant　k　　　 electron rest mass　m_e
　　elementary charge　e　　　　Faraday constant　F

Hartree energy　　E_h　　　　　Planck constant　　h
standard acceleration of free fall　　g_n（数との間は詰めて，$200g_n$ と記す）

　（4）　化学物質名の一部をなすハイフンを伴う元素記号，位置記号 *o*, *m*, *p*, および接頭辞 *sec*, *tert*, *cis*, *trans*, *syn*, *erythro*, *threo* などはイタリック体とする．

　　N-acetyl group　　　　*tert*-butyl group　　　　*m*-dichlorobenzene

　（5）　ローマン体の接頭辞 iso は，イタリック体の *i*- と略記する．

　　isobuty1　　　　　　　*i*-Bu

　（6）　**関数**にはイタリック体を用いるものがある．$f(x)$, $K(T)$

　（7）　**座標軸**や**結晶軸**にはイタリック体を用いる．ハイフンは使用しない．例えば，*x* axis，*a* axis，したがって，直交座標は *xyz* coordinates

　（8）　ベクトル，テンソル，行列，行列式の各成分にはイタリック体を用いる．

　（9）　**対称操作**，**点群**，**空間群**にはイタリック体を用いる．ただし，数字はローマン体である．

　　C_1　　　$C_{\infty v}$　　　D_{3d}　　　O_h　　　$mm2$　　　$P\bar{1}$

　（10）　質量モル濃度にはイタリック体の *m* を，モル濃度にはローマン体の M を用いる．

　（11）　pK_a の p はローマン体の小文字，K はイタリック体の大文字，下付きの a はローマン体の小文字とする．なお，pH の H はローマン体である．複数形はそれぞれ pK_a's，pHs である〔☞ abbreviation（12）〕．

　（12）　強調したい一語または一節をイタリック体で記してもよい．

　（13）　新しい用語を定義したときには，本文中で一回だけイタリック体で記す．同じ目的で引用符も用いられる〔☞ quotation mark（3）〕．

　（14）　動植物，微生物の属名と種名にはイタリック体を用いる．属名の頭字は大文字とし，種名はすべて小文字とする．後者は表題においても，大文字とはしない．ただし，普通名詞の単数形，複数形ならびに形容詞に用いた場合は，小文字のローマン体とする〔☞ capital letter（13）〕．

　　Staphylococcus aureus　　　　*Polygonum* spp.
　　streptococcus　　　　　　　　streptococcal

　（15）　references（and notes）中の論文誌（逐次刊行物）および単行本（書籍）の名にはイタリック体を用いる．前者の巻数もイタリック体とする．

L

Latin term

（1） 普通に用いられる**ラテン語**はローマン体とする．

    ab initio   ad hoc   a priori
    in situ   in vitro   in vivo
    status quo   via   vice versa

これらのラテン語は形容詞として使用する場合には，ハイフンを用いない〔☞ hyphen（6）〕．

    ab initio calculation   in vitro nitrogen fixation
    in vivo nitrogen fixation

（2） ラテン語の略語もイタリック体にしない．いずれも終止符が使用される．次に示すラテン語の意味と使用上の注意は，それぞれの項目または別表の略語，記号を見よ．

    a.m.（ante meridiem）
    ca.（circa）
    cf.（confer）
    e.g.（exempli gratia）   〔☞ e.g.〕
    et al.（et aliti）
    etc.（et cetera）   〔☞ etc.〕
    i.e.（id est）   〔☞ i.e.〕
    p.m.（post meridiem）
    vs.（versus）   〔☞ versus, vs.〕

（3） 出典を記載するにあたって，ibid. および idem は使用しない〔☞ reference（4）〕．

less

little の比較級である．**より少ない**，**より少なく**の意味で量や程度に関する形容詞，副詞のいずれにも用いられる．

  ▶ Potassium perchlorate is much <u>less soluble</u> in a solution containing either another potassium salt or another perchlorate than in pure water.

  ▶ A <u>less generally</u> recognized property of very acidic media is their ability to stabilize monoatomic cations of transition metals in unusually low oxidation states.

なお，数と単位の組合せは集合名詞として扱われるから〔☞ units (22)〕，これに用いる比較級は量を対象とする less である．

 less than 5 mg   less than 3 days

lie in
原因や根拠などが…にあるの表現に用いられる．
- ▶ The source of these discrepancies lies in the use of the hard-sphere kinetic theory of gases in order to evaluate the frequency of collisions.
- ▶ The reason for this lies in the fact that once an oxide of the alkali metals and most of the alkaline-earth metals is produced it reacts further with water with considerable evolution of heat; even the solid hydroxide is not always stable in the presence of water and may go into solution as hydrated ions.

like
…のような，らしいの意の接尾辞で，ハイフンを用いる例が多い．
(1) l でおわる語の後
 bell-like   gel-like   shell-like
(2) 三音節以上の語の後
 resonance-like   radical-like
(3) 化学物質名の後
 adamantane-like   morphine-like
 olefin-like   salt-like
(4) 大文字ではじまる固有名詞および形容詞の後
 Claisen-like
(5) 既にハイフンが用いられている二語からなる形容詞の後
 rare-earth-like   transition-metal-like
 first-order-like

なお，like は**…のような**の意味の形容詞，または副詞として用いられる〔☞ similar (2)〕．

M

major
より大きい，**より重要な**を意味する形容詞で，比較級はない．
- Water is a major constituent of our bodies and of the environment in which we live.
- A major shortcoming of the Langmuir–Blodgett (LB) technique is, therefore, the lack of long-term chemical, thermal, and mechanical stability of the films.

majority
(1) **大多数**，**大部分**の意味の集合名詞である．一つの集合と見なす場合には，動詞は単数形，個々のものを強調する場合には，複数形を用いる．
- Almost all organic compounds can be prepared in the crystalline state, and a majority are crystalline under ambient conditions.

(2) **大多数の**の意味の形容詞として，他の名詞の修飾に用いられる．
- For example, in MgO, the intrinsic thermally produced majority disorder consists of equal numbers of cation and anion vacancies known as Schottky defects.

make
させるの意味の例を次に示す．多くの場合，形容詞が後に続く．
- Thus hydrogen overvoltage makes possible, for instance, zinc and nickel plating and the electrorefining of iron.
- Carbon–carbon cleavage, a key step in the haloform mechanism, is made possible by the presence of three electronegative halogen atoms which inductively stabilize the developing carbanion.
- Such intimate details can be obtained even for quite large molecules, and it is this aspect which makes the fine splittings of NMR spectra of great value in molecular-structure studies.

後に続く語が不定詞の場合，受動態では to を必要とするが，能動態ではこれを欠いてよい．
- At least in principle, any process that tends to proceed spontaneously can be made to do useful work.

mass number
質量数は元素記号の左上につける．これは同位体について記述するときにのみ行う．読み方については，a, an (6) を見よ．

mass spectrometry
質量分光測定の結果の記載方法は，次の例のように一通りではない．
　　MS (EI) m/z: 670 (100, $M^+ - SiMe_3$), 242 (44), 73 (86).
　　ESIMS (m/z): M^+ 1712.
　　HRMS (EI) m/z: M^+ calcd for $C_{17}H_{35}NOSi_4$, 381.1796; found, 381.1789.
　負号－の前後はあける．m/z は質量 m と電荷数 z の比，M^+ は分子イオン，かっこ内の数値は相対強度を表す．一電子移行は矢印 → で表す．

matter
(1)　物質の他に，**事柄**の意味がある．
▶ One <u>matter of</u> special interest, especially in relation to some of the state-to-state studies, is what happens to energy that is in excess of that required to surmount the energy barrier.
▶ A second <u>matter</u> to consider is to what extent the kinetic behavior is affected by interactions between adsorbed molecules, that is, by induced heterogeneity.

(2)　**…の問題**を意味する a matter of ... の形で多く用いられる．
▶ It is a <u>matter of</u> common knowledge that at the melting point, a solid and a liquid can exist in equilibrium with each other in any proportions, as can a liquid and vapor at the boiling point.
▶ The entire question of oxidation state is an arbitrary one, and the assignment of appropriate oxidation states is often merely a <u>matter of</u> convenience.

may, might
主として可能性に関する表現に用いられる助動詞である．
(1)　can よりやや控えめな**…できる**の意味で用いられることが多い．
▶ These primary compounds may then react further, making the total reaction quite complicated. Initial rates <u>may</u> be determined, however, by getting the slope of the pressure–time curve at zero time.

(2)　**もしかしたら…であろう**程度の弱い可能性を表すのに用いられる．
▶ Since one electron is promoted from one definite level to another, we <u>might</u>

expect a sharp spectral line. In fact, we see a moderately broad peak, which is not even symmetrical.

(3)　might は，仮定法の結論部に用いられる〔☞ if ... were〕．

means

(1)　**手段，方法**を意味する場合，単数としても複数としても扱われる．

▶ The combination of IR spectroscopy in the carbonyl region and ^{13}C NMR provides a valuable means of determining the structures of organometallic compounds.

(2)　by means of は**…によって**の意味である．

▶ By means of these complexes good separation of isomers such as xylenes, cymenes, methylnaphthalenes and others has been realized.

(3)　by no means は**決して…でない**の意味である．

▶ Metal alkyls are by no means the only compounds which, on pyrolysis, give rise to free radicals of short life.

▶ The discovery of polonium by Marie Curie in 1898 was the first time, though by no means the last, that invisible quantities of a new element had been identified, separated, and investigated solely by means of its radioactivity.

mechanics

力学に用いる物理量の記号と SI 単位については付表 6 を見よ．

melting and boiling points

融点は mp，**沸点** bp，文献は lit，分解 (decomposition) は dec と略記して，次のように記載する．なお，decompose は decomp と略記する．

　　mp 165.5℃ (lit 165–166℃)

　　mp 220℃ dec

　　bp 135℃

-meter

測定のための機器の意味の接尾辞で，名詞を作る．

　　ammeter　　　　　electrometer
　　photometer　　　　viscometer

-metry

測定法の意味の接尾辞で，不可算名詞を作る．

　　calorimetry　　　　diffractometry

spectrometry　　　　　　　spectrophotometry

mid
（1）　**中間部分の**の意味の接頭辞で，一般にハイフンは用いない．
　　midafternoon　　　　　midcourse　　　　midpoint
（2）　既に接頭辞を含む語の場合は，ハイフンを用いる．
　　mid-infrared

Miller indices
crystal plane and direction を見よ．

molecular weight
分子量はローマン体の小文字を用い mol wt または大文字を用い MW と略記される．用語として**相対分子質量** relative molecular mass が推奨されているので，物理量としての記号 M_r の使用が望ましい．なお，
数平均分子量 number-average molecular weight の記号は M_n，
重量平均分子量 weight-average molecular weight の記号は M_w である．
式量 formula weight はローマン体の小文字を用いて fw と略記される．

month
この綴りは略さない．各月の名は，本文中では略記しないが，脚注，図表の説明，引用文献のリストでは，次のように記載する．

　　Jan　　Feb　　March　　April　　May　　June
　　July　　Aug　　Sept　　Oct　　Nov　　Dec

more than one
一つより多いの意味では複数であるが，続く動詞は単数形を用いる．
▶ When more than one phase is formed from single crystals, the regions occupied by each may be macroscopically visible, or they may range in size down to intimate mixtures in which the individual phases are detectable only by X-rays.

-morph
…**な形態をしたもの**の意味の接尾辞で，名詞を作る．
　　isomorph　　　　　polymorph　　　　　pseudomorph

-morphic
…**の形態をもつ**の意味の接尾辞で，形容詞を作る．
　　dimorphic　　　　　mesomorphic

-morphism
…形態の意味の接尾辞で，不加算名詞を作る．
    dimorphism   isomorphism

morphology
形態または形態学の意味で，後者は不加算名詞である．

most
(1) **たいていの**の意味の形容詞として，冠詞なしで用いる．
- Most inorganic acids generate their protons by rupturing an O–H bond: common examples are oxyacids of nonmetals, such as sulfuric and hypochloric acids.
- Since most lanthanides behave very like lanthanum in most of their reactions, it is not surprising that they occur together in nature and that it is very difficult to separate one from another.

(2) 代名詞として用いた most of ... の形は量**の大部分**を意味する．動詞は単数形とする．
- Liquid zinc is insoluble in liquid lead, and the solubility of silver in liquid zinc is about 3000 times as great as in liquid lead. Hence, most of the silver dissolves in the zinc.

much of ...
不可算名詞につける．**量**または**程度**を表し，動詞は単数形を用いる．
- Much of our knowledge of organic chemistry has resulted from investigations of the composition of naturally occurring substances.
- Because of this lack of appreciation of Gibb's work, much of the same ground was covered independently by others.

なお，数を表すときには many of ... を用い，可算名詞の複数形につける．

multi
多数の，**多種の**の意味の接頭辞で，一般にハイフンは用いない．
    multicolored   multidentate   multifunctional
    multilayer   multiphoton   multiple

N

namely

すなわち，**言い換えれば**の意味の副詞で，that is, i.e. と同様に，より正確な情報をつけ加えるときに用られる．二つの節の間では，前をセミコロン，後をコンマで切る．それ以外では，前後ともにコンマを用いる．

▶ Using different reaction conditions, namely, vapor-phase reaction of iodine with pure C_{60} at 250°C for several days in evacuated Pyrex tubes, we readily obtained a highly crystalline phase of ideal stoichiometry $C_{60}I_4$.

neither

（1） not either すなわち，**どちらも…でない**の意味であるから，二つのものを対象とするのが原則である．

▶ While orientation and induction effects are undoubtedly relevant in solids in which molecules are polar, neither can play any part in the numerous structures, such as those of the inert gases, methane, hydrogen, benzene and many others, in which the atoms or molecules possess no permanent dipole moment.

（2） neither of の次の名詞が複数形であっても，動詞は単数形を用いる．

▶ Neither of these extreme situations is at all likely.

（3） neither A nor B が主語の場合，動詞の数は B に一致させる．

▶ In most chemical processes neither the energy nor entropy is held constant.

▶ A system is said to be isolated if neither matter nor heat is permitted to exchange across the boundary.

A と B は文法上，対等の関係にあればよく，名詞に限らない．

▶ Band splitting is always observed for xanthenes, indicating that these aggregates are neither parallel nor linear but have intermediate geometry.

▶ Methylamine neither gains nor loses resonance stabilization in going from the free base to the conjugate acid.

nevertheless

それにもかかわらずの意味の副詞で，in spite of, however, yet と同じ意味に用いられる．コンマ，セミコロンの扱い方は however の場合に同じ．

▶ <u>Nevertheless</u>, hydrogen appears to migrate more selectively than expected if the hydrogenation agent were a free H atom.

▶ Ionic interactions are considerably stronger than hydrogen bonds, and they usually dominate a packing pattern; <u>nevertheless</u>, hydrogen bonds may still be useful as organizing forces even in the presence of ionic interactions.

newly, recently

ともに**最近**の意味で使用されるが，前者は過去分詞と組み合わせて用いられる．複合形容詞となっても，ハイフンは使用しない〔☞ hyphen (5)〕．後者は文中の強調したい箇所におくことができる．

NMR spectroscopy

(1) **核磁気共鳴分光**に用いる物理量の記号と SI 単位については，付表 7 を見よ．

(2) 幾通りもの測定結果の記載方法が行われる．場合に応じて選択する．

^{1}H NMR(500 MHz, CDCl$_3$, δ) : 0.88 (t, J = 6.78 Hz, 3H), 1.25 (t, J = 7.33 Hz, 3H), 1.26−1.41 (m, 7H), 2.36−2.46 (m, 3H), 2.50−2.58 (m,1H), 2.64−2.72 (m, 1H), 2.80−2.88 (m, 3H), 4.14 (q, J = 6.78 Hz,2H), 7.43 (s, 1H).

^{1}H NMR (Cl$_2$CDCDCl$_2$, 500 MHz) : δ 1.23 (s, CH_3), 1.43 (s, CH_3), 1.96 (s, β-CH_2), 2.07 (s, β-CH_2), 2.40 (s, β-CH_2), 2.52 (s, β-CH_2), 4.90 (s, α-CH_2), 5.11 (s, α-CH_2), 5.42 (s, α-CH_2), 6.75 (s, α-CH_2), 7.20 (s, methine), 14.85 (s, pyrrole).

^{13}C NMR (CDCl$_3$, 125 MHz) : δ 14.04, 14.26, 22.54, 26.27, 28.16, 30.64, 34.79, 34.94, 45.53, 60.62, 131.96, 138.54, 172.38, 202.13.

δ は ppm 単位で表した化学シフトで，外部磁場に比例し，溶媒によって変動する．基準物質 TMS の吸収に対して，ほとんどの場合に負の値となるが，便宜的に符号を逆にして，正の値として用いる．スピン-スピン結合定数 J は外部磁場の強さに無関係である．シフトに関与している元素または基はイタリック体で表示する．多重度は，s；singlet, d, doublet；t, triplet；q, quartet と略記する．

no

一つ（一人）もないの意味で，名詞の単数形，複数形いずれの前に用いてもよい．用いる動詞が単数形か複数形かは，名詞の数によって決まる．ただし，no data は単数形として扱われる〔☞ data〕．

▶ No single value of the atomic refraction of fluorine will, when combined with the standard refractions of other atoms, produce good agreement with the observed molar refractions of all fluorine-containing compounds.

▶ Before our work, there were no examples of stable species at room temperature containing covalent S–I or Se–I bonds except SeI$_6$.

non-

名詞，形容詞，副詞につける**無**，**非**，**不**など，否定や欠如の意味を表す接頭辞である．

(1)　一般にハイフンなしで続けられる．

noncombustible　　　noncovalent　　　　　　nonlinear
nonpolar　　　　　　nonstoichiometric　　　　nonvolatile

(2)　Webster 系の辞典では，n ではじまる語でもハイフンは用いない．

nonnative　　　　　nonnatural

(3)　化学物質名の前ではハイフンを使用する．

non-hydrogen bonding　　　non-phenyl atoms
non-alkane

(4)　固有名詞とそれに由来する形容詞の前ではハイフンを用いる．

non-Coulombic　　　　　　non-English-speaking
non-Gaussian　　　　　　　non-Newtonian

(5)　既に接頭辞を含む語にはハイフンを用いる〔☞ prefix (2)〕．

non-superconducting copper oxides

(6)　既にハイフンを用いてある形容詞句には，さらにハイフンを用いる．

non-diffusion-controlled system
non-radiation-caused effects
non-tumor-bearing organ

none

この不定代名詞は意味によって単数形とも，複数形ともみなされる．

▶ None of the three elements (arsenic, antimony, and bismuth) is particularly abundant in the earth's crust though several minerals contain them as major constituents.

▶ None of the phenomena mentioned so far are restricted to titania.

nonetheless
それにもかかわらずの意味の副詞である．
- ▶ The chemistry of vanadium, although too rich, complex, and colorful to be predictable, seems, <u>nonetheless</u>, to be logically understandable in terms of electronic structure.

nor
否定を表現する叙述または節を結ぶのに用いられる接続詞である．
- ▶ Mercury does not tarnish in air <u>nor</u> combine appreciably with oxygen below 350 ℃.

not only ... but, not only ... but also
(1) **A だけでなくまた B** を意味する not only A but (also) B が主語である場合，動詞の数は B に一致させる．
(2) A と B は名詞であるとは限らないが，文法上，対等の関係にあることを要する．
- ▶ Water forms clathrates <u>not only</u> with the noble gases <u>but</u> with a wide range of other molecules.
- ▶ Conformational analysis can account <u>not only</u> for the fact that one conformation is more stable than another <u>but</u> often for just how much more stable it is.
- ▶ The contribution of each C–C and C–H bond is supposed to be influenced <u>not only</u> by its nearest neighbor bonds <u>but also</u> by neighbors once, twice, and further removed.

nothing other than
…にほかならないの意味で，nothing but も同じように用いられる．
- ▶ The potential-energy surface for such a reaction is <u>nothing other than</u> the common potential-energy curve for a diatomic molecule.

noun
名詞には数えられる名詞，すなわち**可算名詞**（普通名詞，集合名詞）と，数えられない名詞，すなわち**不可算名詞**（物質名詞，抽象名詞）がある．ただし，一つの名詞でも，複数の語義があって，それぞれ可算名詞，不可算名詞と異なることもあるので注意を要する．

(1) **普通名詞**

atom	complex	compound
decrease	deviation	galvanometer
molecule	phase	pressure
solid	solution	state
substance	surface	volume

(2) 集合体を1単位として見る**集合名詞**

couple	group	majority
number	pair	series
staff	team	variety

一つの集団と見なすときには，動詞の単数形と組み合わせ，集団を構成する個々を意味するときには，動詞の複数形と組み合わせる．可算名詞である集合名詞の前に数をつけるときには注意を要する．例えば，three staff は不可で，three of the staff または three members of the staff が用いられる〔☞ majority, series, variety〕．

(3) **物質名詞**

copper	glass	light
paper	steam	water

(4) **抽象名詞**

assistance	chemistry	diffractometry
height	length	means
polarography	principle	time

不可算名詞には複数形はなく，不定冠詞はつかないし，一般的なことを述べるときには，定冠詞を必要としない〔☞ the (4)〕．

▶ Palladium is another rare metal that occurs within platinum and is extracted from its residues.

▶ Infrared spectroscopy is the observation of the absorption of infrared radiation by the vibrational modes of molecules.

特定されたものの場合には，定冠詞または所有格をつける．

(5) 名詞（特に抽象名詞に多い）には特定の前置詞を伴うものがあるから，辞典を参照することが望ましい．

　　(a) to を伴うもの： approach, contribution, exception, introduction, reference, reply, solution など

（b）　for を伴うもの：　demand, desire, need, regard, respect, room, substitute, synonym など

　（c）　on または upon を伴うもの：　attack, comment, dependence, effect, reflection, restriction など

　（d）　with または between を伴うもの：　affinity, identification など

　（e）　その他のもの：　control over, departure from, quotation from など

number, numeral

（1）　**数字**を文頭に用いることは避けて，数と単位の両方ともを綴る．

　　Twenty-three milliliters of ethanol was added, then ....

あるいは，次のような表現を用いる．

　　Ethanol (23 mL) was added, then ....

　　A 23-mL portion of ethanol was added, then ....

　量または時間に関する主語が複数形の名詞であっても，動詞は単数形を用いることに注意せよ．これらの場合でも，文中では数と単位の組合せを用いて差し支えない．

　▶Twenty-three milliliters of ethanol and 40 mL of acetic acid were mixed.

（2）　数字と単位の間はあける．

　　5 min　　　30 mL　　　100 g　　　　　120 ℃

ただし，%，°，′，″ は例外である．

　　48%　　　60°30′　　　30′15″

（3）　四桁まではコンマで区切ることはしない．五桁以上の数はコンマで区切る．ただし，表の同じ列に四桁と五桁以上の数がある場合は，両者ともにコンマを使用する．

（4）　測定の精度に応じて，小数やゼロを用いる．

（5）　単位がついているときに，分数は用いない．小数にはコンマではなく，小数点を用いる．小数点の前後には数字がなければならない．25.0 ℃と25 ℃は同じではない．

（6）　標準偏差や誤差は，数値に続けてかっこ内に示すか，±に続けて桁を明らかにして示す．記号±の前後はあける．

　　2.0189 (11)　　または　　2.0189 ± 0.0011

（7）　測定単位を伴う大きな桁の数は，適当な単位を採用するか，$\times 10^n$ を用いて，四桁以上の数は避けた表現を使用する．

23,400 mL ではなく 23.4 L
1.23×10^4 kg　　　1.5×10^5 s　　　8.5×10^{-4} M

ただし，表の中では同じ単位を用いる．このため，一部に四桁以上の数を生じることがあってもやむを得ない．

(8) 次の単位は，四桁以上の数になっても，使用することが望ましい．
　　mass density of fluids g L^{-1}
　　mass density of solids kg m^{-3}
　　modulus of elasticity GPa
　　fluid pressure kPa
　　stress MPa

(9) 数と単位の記号からなる形容詞にはハイフンを使用する．二つの数の間に二分ダッシュまたは to が用いられているときも同様である．
　　10-min exposure　　　　　20-mg sample
　　a 20–25-mL aliquot
　　a 1- to 2-h sampling time　　　a 20- to 25-mL aliquot

次の例外がある．
　(a)　単位が M, *m*, mol dm^{-3}, mol % ときにはハイフンを用いない．
　(b)　温度の単位，℃, K にはハイフンを用いない．
　(c)　単位が °, ′, ″, % のときには，数との間を詰めるので，ハイフンを用いない．
　(d)　$\times 10^n$ をつけた数値にはハイフンを用いない．

(10) 時間や測定単位がついていないとき，数学的表現ではないときには，9 までの数は綴り，10 以上は数字を用いる．
　　three test tubes　　　　　20 test tubes
　　ninefold　　　　　　　　10-fold
　　fifth example　　　　　　six times

ただし，系列や範囲を表す数詞，対等の関係にある数詞に 10 以上を含む場合は，9 までの数にも数字を用いる．このとき，二分ダッシュの前後は詰める．
　　6–15 times

(11) 数学的表現では，常に数字を用いる．
　　by a factor of 2　　　　　3 orders of magnitude

5 times

(12) 比は数字と：または / を用いて表す．いずれも前後は詰める．
　　 a ratio of 1:10　　　　　　a ratio of 1/10
　　 a 1:2 (v/v) mixture　　　　a 1/2 (v/v) mixture

(13) 分数は分母子ともに 10 以下の場合は綴り，ハイフンを使用する．
　　 one-half　　　　　　　　　one-quarter
　　 two-thirds　　　　　　　　three-fourths
分母子のいづれかが 10 以上の場合は，1/10, 13/20 などで表現する．

(14) 比を表すときには，常に数字を用いる．
　　 a ratio of 1:2　　　　　　 a 1/1 (v/v) mixture

(15) 年代を表すには，'s ではなく s をつける．すなわち，1990's ではなく，1990s とする．

(16) 日付には数字を用いる．April 1st ではなく April 1 とする．

(17) a.m., p.m. には数字を組み合わせる．
　　 12:10 a.m.　　　　　　　　6:00 p.m.

(18) figures, tables, schemes, structures, equations, references には番号をつける．引用文献にはアラビア数字を用いるが，それ以外にはローマ数字を用いることもある〔☞ chemical equation (1)〕．

(19) **化合物番号**には，ゴシック体のアラビア数字を用いる．
　　 (a) 正式名の後には，かっこ ( ) に入れた番号をつける．
　　 (1,2,3-η-2-butenyl)tricarbonylcobalt(I) (**1**)
　　 2,3-quinolinediacetic acid (**2**)
　　 poly[formaldehyde-*alt*-bis(ethylene oxide)] (**3**)
　　 (b) 化合物の正式名が記載されていないとき，あるいは既に番号が与えられた化合物は，かっこなしの番号を用いる．
　　 keto alcohol **4**
　　 compounds **2**, **5**–**8**, and **10**

(20) 本文中の節や文に番号をつけるときには，かっこ ( ) に入れた数字を用いる．

(21) 図表，化合物，引用文のそれぞれ二つを指定するには and を用いる．
　　 Figures 1 and 2　　　　　　Tables 3 and 6
ただし，数字だけがかっこ内にあるとき，上付き文字のときには，コンマを

用いる．後者の場合はコンマの後を詰める〔☞ citation in text〕．

(22) 三つ以上である範囲を占める場合は，上付き文字の場合も含めて，二分ダッシュを用いる．二分ダッシュの前後は詰める．

  compounds **1**–**3**   refs 13–16
  Corey (11–13)   Corey$^{11-13}$

from ... to, between ... and の表現の一部として，二分ダッシュを用いてはいけない．

(23) 数字の前に形容詞の働きをしている演算子がある場合には，二分ダッシュでなく，to を用いる．

  −15 to +100 ℃   20 to >500 mL

(24) × 10^n を用いて表した数の範囲は，省略なく記載するか，かっこを用いて誤解を避ける．

  8.8×10^{-3} to 12.4×10^{-3} または $(8.8–12.4) \times 10^{-3}$

number of

(1) a number of, a large number of は**多数の**を意味し，その後の名詞は複数形，動詞も複数形を用いる．ACS では，a number of よりは，簡単な many または several の表現を用いることが望ましいとする．

▶ A number of theoretical investigations have been made to account for the solvent effect on the measured electric moment.

▶ There are a number of ways to achieve unsaturation in a molecule.

(2) 主語である number が実際に数を意味するときには，名詞の単数形として扱われ，その前には the, this または that がつく．the large number となっても，変わりはない．

▶ The large number of valence orbitals in polynuclear metal–metal-bonded complexes, coupled with their highly delocalized character, leads to a breakdown of simple orbital descriptions of ground- and excited-state reactivities.

O

obtain
get, obtain を見よ．

obvious
明白なを意味する形容詞で，**目につきやすい**の意味をもつ．
- ▶ The heterogeneity of a piece of granite is obvious to the eye.
- ▶ The most obvious explanation of the facts concerning passivity is that it is due to a protecting film.

of
of を用いた次の例文は形式的に類似しているが，意味はまったく異なる．
(1) 物質，材料などを示す．ただし，材料の質に変化は生じない．
- ▶ The balance is made of quartz; it pivots on tungsten needle points borne in quartz cups; and its oscillations are followed with the aid of an optical lever, reflected from mirror to a scale outside the air thermostat in which the balance is housed.

(2) …**に関して**の意味で用いる．
- ▶ A considerable number of measurements have been made of the enthalpy and entropy changes that occur on adsorption and desorption.

on
(1) on と upon はしばしば交換できるが，on は位置や静止状態を，upon は方向や動きを強調する．前置詞 onto は upon と同様に方向や動きを示唆する．

(2) 受動的表現において，測定機器を表す前置詞として用いる．
- ▶ Infrared spectra were recorded on a Nicolet 7199 Fourier transform spectrometer.

recorded の代わりに acquired, obtained, run も用いられる．

on ...ing
…**すると**の意味を表す．in ...ing と比較せよ．
- ▶ On cooling, the nitric oxide is further oxidized to nitrogen dioxide.

すなわち，when the nitrogen oxide is cooled を意味する．

- ▶ <u>On heating</u> the Fe^{2+}-containing minerals in air, oxidation of Fe^{2+} to Fe^{3+} occurs first.
- ▶ <u>On multiplying</u> the left sides and the right sides of these equations, we obtain ....

one of

名詞の複数形の前にあるときも，主語は単数であるから，動詞は単数形を用いる．例文中の chemistry は不可算名詞である．

- ▶ <u>One of</u> the major differences between organic and inorganic chemistry is the relative emphasis placed on structure and reactivity.

しかし，**one of those...** ＋関係代名詞の場合には，その次の動詞の数は先行詞 those... で定まるから，複数形となる．

only

(1) 副詞としては，修飾する語，句，節の前または後に置く．

- ▶ The system salt–water is binary <u>only</u> in the absence of hydrolysis.
- ▶ The overall charge distribution in a molecule is a resultant of a number of factors, of which the bond polarity is <u>only</u> one.
- ▶ From stereochemical point of view, nitrogen is second in importance <u>only</u> to carbon.

似た意味で，alone が使用される．

- ▶ Silver forms all four monohalides. The fluoride <u>alone</u> is soluble in water.

(2) the only は**唯一の**の意味の形容詞である．

- ▶ <u>The only</u> isomerism for tetrahedral complexes is optical isomerism.
- ▶ The system $FeCl_3$–H_2O is binary as long as <u>the only</u> solid phases are ice, $FeCl_3$, or hydrates of $FeCl_3$ and as long as the vapor, if present, is pure H_2O or H_2O and $FeCl_3$.

or

(1) 二つまたはそれ以上の選択すべき文法上，対等の関係にある語，句，節をつなぐ．

- ▶ Since the heat of a reaction depends on whether a reagent is solid, liquid, <u>or</u> gas, it is necessary to specify the state of the reagents.
- ▶ If a substance, neither a reactant nor a product, affects the rate it is called an inhibitor, retarder, sensitizer, <u>or</u> a catalyst, depending on the nature of the

effect.

いずれも単数形の名詞ならば，動詞は単数形にする．いずれも複数形の名詞ならば，動詞は複数形とする．もし，or が名詞の単数形と複数形を組み合わせるならば，動詞の形はより近くに位置する主語に合わせる．したがって，one or two の次にくる動詞は，two に一致させた複数形とする．

(2) 次の場合には，接続詞 or を重ねて用いる．
- ▶ Only true chemical substances <u>or</u> species, <u>or</u> definite combinations of such, can be components of a system.
- ▶ The octahedra are then joined up, often with some distortion, by sharing edges <u>or</u> corners, <u>or</u> by almost any combination of these.

(3) **すなわち，言い換えれば**の意味の場合は，コンマで区切った中に類義語（句）をおく．
- ▶ The forces operating in ionic crystals are very predominantly the electrostatic, <u>or ionic</u>, forces between the charged ions.
- ▶ The chemical inertness of alkali halide nanocrystals, based on their composition of closed-shell atomic ions, contrasts sharply to the reactivity of metal or semiconductor clusters which have unsaturated, <u>or dangling</u>, bonds at their surfaces.

(4) 重文における接続詞 or の前には，コンマを用いる〔☞ comma (11)〕．

(5) 接続詞 or の代わりにスラッシュを使用しない．

ordinary, ordinarily

普通の意味をもつ形容詞と**普通に**の意味をもつ副詞である．
- ▶ Thus, although calcite is the most stable form of calcium carbonate at the <u>ordinary</u> temperature, the metastable modification, aragonate, nevertheless exists under the <u>ordinary</u> conditions in an apparently very stable state.
- ▶ Binary systems can <u>ordinarily</u> exist in a number of phases.

other things being equal

他の条件が同じならの意味の慣用句として用いられる．
- ▶ <u>Other things being equal</u>, an exothermic reaction will proceed more readily on a thermodynamic basis than will an endothermic reaction.

owing to

…のためにの意味で，because of, on account of と同義である．

▶ <u>Owing to</u> the small size and extreme electronegativity of fluorine the structural chemistry of this halogen is very different from that of the other members of the family.

▶ The reduction of mixtures of metal oxides and boric oxide with carbon is not generally satisfactory <u>owing to</u> heavy losses of boric oxide by volatilization and contamination of the product with boron, carbon, and boron carbide.

oxidation number

酸化数にはローマ数字を用いる．負号は数字の前に詰めて記す．イオン電荷の場合とは，数字と負号の順が逆である〔☞ ionic charge〕．

(1) 元素記号の右上につける．

Ni^0 $\quad$ Cu^{II} $\quad$ Co^{III} $\quad$ V^{IV} $\quad$ I^{-I} $\quad$ O^{-II}

(2) 複数個の原子から成り立つときには，上付き文字の酸化数，次いで下付き文字の原子数の順序となる．イオン電荷の場合の原子数，電荷数の順序と混同しないよう注意せよ．

Pb^{II}_2 $\quad$ O^{-I}_2 $\quad$ Si^{-I}_4

(3) 元素記号や名称に詰めて続け，かっこ（ ）に入れて記載する．

iron(II) $\quad$ sulfur(VI) $\quad$ hexafluorophosphate(V)

oxidize

(1) 他動詞として**酸化させる**の意味で用いられる．

▶ CrO_3 <u>oxidizes</u> primary alcohols to aldehydes.

▶ When small-ring cyclic glycols <u>are oxidized</u> by lead tetraacetate, the cis isomer is transformed at a greater rate than the trans isomer.

(2) 自動詞として**酸化する**の意味で用いられる．

▶ When heated in air, arsenic sublimes and <u>oxidizes</u> to As_4O_6 with a garlic-like odor.

P

parentheses

かっこには，( )，[ ]，{ }，〈 〉の種類があるが，ここでは，最もよく使用される（ ）（パーレン）について記する．

(1) 完結した本文中，補足説明（略語，記号の定義を含む）があれば，その位置にかっこ（ ）に入れて付記する．

▶ The infrared spectrum of this product (KBr wafer) showed absorption at 3030 cm^{-1} (unsaturated CH), 2940 cm^{-1} (saturated CH), 2250 cm^{-1} (nitrile), and 1612 cm^{-1} (conjugated$-C=C-$).

(2) かっこ内が補足説明文の場合，それが本文の終止符より前にある場合は，かっこ内の文頭には大文字は使用しないし，終止符もつけない．

(3) 本文の終止符の後に補足説明文が独立して存在する場合は，かっこ内の文頭には大文字を用い，終止符もつける．

(4) 同位体置換された化合物名を綴る場合，同位体の記号と個数（必要ならば位置を指定する数字）はかっこ（ ）内にまとめて示す．かっこと化合物名との間は詰める．同位体修飾化合物の化学式については，bracket (5) を見よ．

(^{15}N)ammonia   (1,3-^{3}H$_2$)benzene   chloro(^{3}H)benzene

(5) 具体的な複数個の説明に番号や記号をつける場合は，本文中ならば (1)，(2)，(3) ...，図の説明ならば (a)，(b)，(c) ... と記す．

▶ Similitude effects have likewise been looked for in at least two other types of crystals: (1) when the units of structure are ionic; particularly for salts the electronic structure of whose ions corresponds with that of the inert gases; (2) when the units are metal atoms.

(6) 試薬や機器の製造業者名を付記するときには，これらをかっこ内に記し，大文字ではじめる〔☞ capital letter (11)〕．

(7) 本文中の文献番号は本文の一部である．補足説明ではないから，かっこは用いない．すなわち，ref (1) とはしないで，ref 1 と書く．

この他，化学物質名，化学方程式，数式にも，かっこは使用される〔☞ chemical name, chemical equation, equation〕．

part

(1) **構成要素の一つ**である部分を表す名詞である．
▶ A heterogeneous material is a material that consists of <u>parts</u> with different properties.

(2) part of, a part of は可算名詞または不可算名詞の単数形の前に用いる．複数名詞の前には some of, many of を用いる．
▶ A phase is a homogeneous <u>part of</u> a system, separated from other parts by physical boundaries.

(3) 数詞がつくと，**割合**を表す．
▶ Dalton thought that <u>seven parts</u> by weight of oxygen combine with <u>one part</u> of hydrogen to form water.

(4) in large part は largely と同義である．
▶ The properties of solutions have been extensively studied, and it has been found that they can be correlated <u>in large part</u> by some simple laws.

partially

completely の反対で，**部分的に**はこうだが，全体としては不十分であることを意味する．
▶ Some of the approaches used to describe bonding between atoms that have <u>partially</u> filled s and p orbitals have been described.

particular

同類中，**選別的に特殊**であることを意味する形容詞として用いる．
▶ In spite of this difficulty, many reactions studied kinetically can be explained by a <u>particular</u> set of simple processes which are so reasonable and so in accord with all chemical experience that we accept them as essentially true.

partly

wholly の反対で，全体がどうであるかに関係なく，**部分的に**はこうだ，を意味する．partially と比較せよ．
▶ The failure to isolate the cis complex may be due <u>partly</u> to the probable greater solubility of this isomer and possibly also to greater stability of the trans form.

pass

光が通るの表現に自動詞としても，他動詞（受動態）としても用いられ，方

向を表す表現を伴う．
- ▶ When a ray of light passes from one substance into a different substance, it usually changes its direction of travel.
- ▶ The laws of Lambert and Beer are concerned with the intensities of light absorbed or transmitted when incident light is passed through some material.

past tense, present tense
過去時制と**現在時制**の用法についての注意を以下に記す．
(1)　過去時制は著者ならびに他人が行ったことを述べるときに用いる．
- ▶ Indium was first detected from the indigo line in its spectrum and was named accordingly.
- ▶ The solid mixture was heated in a furnace maintained at 400 °C.

(2)　現在時制は**事実**を記載する場合に用いる．
- ▶ Oxygen is the most abundant element in the earth's surface.
- ▶ An ionic bond is an electrostatic attraction between oppositely charged ions.

(3)　**結果**，**考察**，**結論**を述べるときは，現在時制と過去時制のどちらを用いてもよい．なお，主節の動詞が現在形のときには，従属節の動詞は状況に応じた時制をとればよい．
- ▶ It is only for the hydrogen atom itself that the potential energy curve is exactly $-e^2/r$.
- ▶ It must be emphasized that the hybridization theory was developed to explain the facts.

(4)　主節の動詞が過去形あるいは過去完了形のときには，従属節の動詞は時制の一致を必要とする．
- ▶ Arrhenius postulated that some unspecified active parts of the molecules were formed and were responsible for carrying the electric current.

(5)　従属節が一般的真理，現在も変わらない事実，習慣，性質を表すときには，主節の時制に関係なく，現在時制とする．
- ▶ One important observation Bunsen and Roscoe made was that a small amount of light can bring about the conversion of a great deal of hydrogen and chlorine gases to hydrogen chloride.

peri

(1) **まわりの，まわりに**の意味の接頭辞で，一般にはハイフンは用いず，一語とする．

  pericyclic   perimeter   periphery

(2) **近い，隣の**の意味の場合は，ハイフンを使用する．

  peri-position

period

(1) 文末の**終止符**と次の文頭との間は2字分あける．

(2) 略語，記号の後には，他の語との混乱のない限り，終止符はつけないことが多い〔☞ abbreviation (7)〕．

(3) 終止符をもつ略語，記号，例えば etc. や et al. が文末にあるときには，さらに終止符をつけることはしない．

(4) 終止符を他の句読点 ?, ! と併用しない．

photo

光に関係した，光により生じたの意味の接頭辞で，o が重なる場合を含めて，ハイフンは用いない．

photocatalyst	photochemistry	photocurrent
photoelectron	photograph	photoionization
photolysis	photometry	photon
photooxidation	photooxygenation	photopolymerization
photoreaction	photosynthesis	photovoltaic

physical quantities

物理量の記号については付表 1–12 を見よ〔☞ italic type, Greek letter〕．物理量の記号と数値に間には等号 ＝ を使用する．

(1) 物理量の記号にはイタリック体またはギリシャ文字を使用する．

(2) 物理量の記号は，下付き文字となったときもイタリック体とする．

  C_P $C_{V,\mathrm{m}}$ ΔS_m

(3) C_P の複数形は C_P values または C_P's とする．

(4) 単位記号はローマン体とし，units に示された SI 単位を使用する．

polymer chemistry

高分子化学に用いる物理量の記号と SI 単位については付表 8 を見よ．

portion
part と同じように用いられるが，**割り当てられた部分，切り取られた一つ**の意味の部分である．
- ▶ Any portion of matter on the earth is attracted toward the center of the earth by the force of gravity; this attraction is called the weight of the portion of matter.
- ▶ Although it is a relatively simple experimental matter to measure the internuclear distances in ionic crystals, the determination of that portion of the internuclear distance contributed by the cation and that portion by the anion is not so simple or even obvious.

post-
時間的に**後の**の意味の接頭辞である．
(1) 一般にはハイフンを用いないで一語とする．
 postexposure   postpolymerization
 postvulcanization
(2) 既に接頭辞を含むか，t が重なる場合にはハイフンを用いる．
 post-reorgnization   post-translational

pre-
前の意味の接頭辞である．
(1) 一般にはハイフンを用いないで一語とする．
 precooled   predissociation   presetting
(2) e ではじまる語の前，既にハイフンで結ばれている二語の前ではハイフンを要する．
 pre-equilibrium   pre-exponential
 pre-steady-state condition
ただし，例外も多い．
 preexperimental   preexposure

prefix
(1) 次の**接頭辞**（連結形と呼ばれるものを含む）はハイフンをつけないで用いる．

a, an	bio	carbo	chalco
chromo	cryo	di	electro

endo	equi	exo	extra
galvano	hydro	hyper	hypo
in, im, ir	intro	iso	lyo
lyso	macro	mega	meso
meta	metallo	mini	mono
multi	nano	neo	ortho
over	para	photo	physico
piezo	poly	proto	pyro
radio	retro	semi	stereo
super	syn	techno	thermo
trans	tri	ultra	under
uni	up	video	visco

英国系の辞典には，米国系の辞典よりもはるかに多くのハイフンが使用されている．米式の英語で論文を書くには，使用する辞典に配慮を要する．

(2) 次の接頭辞では，例外的にハイフンを使用することがある．それぞれの項目を参照せよ．

anti	auto	bi	by
co	counter	hetero	homo
inter	intra	mid	non
peri	post	pre	pseudo
re	sub	un	

その理由として，ACS には次の三つが示されているが，それですべてが説明できる訳でもないので，Webster 系の辞典を参照することが望ましい．

(a) 同じ文字が重なる場合，例えば

  anti-infective   inter-ring
  post-translational   sub-bandwidth

にハイフンを用いるというが，例外も多い．

  cooperation   coordination
  nonnegative   nonnitrogenous
  nonnuclear   preelectrolysis
  preexperimental   preexposure
  unnamed   unnecessary

(b) 既に接頭辞を含む語の場合にはハイフンを用いる．
　　anti-inflammatory　　　anti-overshooting
　　bi-univalent　　　　　　mid-infrared
　　non-superconducting　　post-reorganization

(c) 綴りが同じで，意味が異なる語が存在するとき，ハイフンを使用して区別できるようにする．

recollect と re-collect	**思い出す**と**再び集める**
recover と re-cover	**取り戻す**と**再び覆う**（または**張り替える**）
reform と re-form	**改良する**と**作り直す**
unionized と un-ionized	**連合した**と**イオン化していない**

(3) **倍数接頭辞**は名称の前につけ，間隔をあけず，ハイフンも用いない．一般に倍数接頭辞の終わりの母音は削除しない．
　　triiodide　　　　tetraalkylammonium　　　　pentaoxide
ただし，monoxide は例外で，monooxide とはしない．

(4) ギリシャ語とラテン語に由来する倍数接頭辞は区別して使用する．
　(a) ギリシャ語に由来する倍数接頭辞
　　　hemi　　mono　　di　　tri　　tetra　　penta　　hexa
　　　hepta　　octa　　nona　　deca　　undeca　　dodeca, ...
　　　bis　　tris　　tetrakis　　pentakis　　hexakis　　heptakis
　　　octakis　　nonakis　　decakis, ...
　(b) ラテン語に由来する倍数接頭辞
　　　semi　　uni　　bi　　ter　　quater　　quinque
　　　sexi　　septi　　octi　　novi　　deci

(5) ハイフンを用いた形容詞に，さらに接頭辞をつける場合には，ハイフンを使用する〔☞ hyphen (11)〕．
　　non-diffusion-controlled system
　　pre-steady-state condition

(6) 化学物質名に接頭辞をつけるときにはハイフンを使用する．
　　non-hydrogen bonding　　　non-phenyl atoms
　　non-alkane

(7) 数字に接頭辞をつけるときにはハイフンを用いる．
　　pre-2000s

(8) 固有名詞とそれに由来する形容詞に接頭辞をつけるときは，大文字をそのままにして，ハイフンを使用する．

    anti-Markovnikov   non-Gaussian

    non-Newtonian

(9) 複数の接頭辞が，対等の関係で後続する語に帰属するときは，ハイフンを使用して簡略化できる〔☞ hyphen (19)〕．

    pyroelectric and piezoelectric materials は

    pyro- and piezoelectric materials とする．

proportion

比率の意味を含む部分を表す．

▶ X-ray diffraction patterns obtained from bulk polymers are diffuse and exhibit a considerable <u>proportion</u> of background scattering, revealing the presence both of small crystallites and of regions in which the molecules are severely disordered.

provide, provided that

(1) provide は**提供する**の意味の他動詞である．

▶ A good example <u>is provided</u> by the reaction between triethylamine and ethyl iodide, studied many years ago by Menschutkin in 22 different solvents.

(2) provided は**…という条件で，もし…とすれば**の意味の接続詞で，しばしば provided that の形で用いられる．

▶ Silica glass fiber, <u>provided that</u> its surface is protected as noted previously, can be made sufficiently perfect and flaw-free that kilometer lengths can withstand an extension of 1 percent or more.

pseudo-

(1) **疑似の**の意味を表す接頭辞で，一般にはハイフンは用いない．

    pseudomorph   pseudopotential

(2) 既にハイフンで結ばれている二語の前ではハイフンを要する．

    pseudo-first-order reaction

Q

quantitative analysis

定量分析の結果は，コロン，セミコロン，コンマを使用して，次の例のように記載する．

▶ Anal. Calcd for $C_{21}H_{20}N_8O_9$: C, 47.75; H, 3.81; N, 21.24%. Found: C, 47.69; H, 3.94; N, 20.98%.

Found と Calcd の記載順序，%を使用するか否かは論文誌によっても異なる．

quotation mark

引用符 " " の使用では，コンマや終止符が引用の一部であるか否かで，二番目の引用符" の位置が定まることに注意せよ〔☞ comma (14)〕．

(1) 本文中に論文，単行本，章，政府刊行物の題名を引用するときは，これを引用符で囲む．

▶ The title of Slater's paper, "The Self-Consistent Field and the Structure of Atoms," shows his debt to Hartree, although Slater's method turned out to be a great deal more practical than Hartree's, as well as consistent with the methods of Heitler, London, and Pauling.

(2) 50語を超えない引用文は引用符で囲む．引用符 " は終止符の後になる．引用するのは文に限らず，語や節でもよい．

▶ There is an old saying that "physical chemistry is the study of all those quantities whose negative logarithm is linear with $1/T$."

▶ Coulson defined the meaning of "fractional bond order" for bonds intermediate between integral values, at last putting Thiele's old idea of "partial valence" on firm theoretical footing.

(3) 本文中，新しい意味で用いた語は，最初の一回だけ引用符で囲む．イタリック体を用いてもよい〔☞ italic type (13)〕．

▶ "Convenience foods" for both humans and their pets, prepared mixes for cakes and other baked goods, instant coffee, and frozen foods all gained popularity throughout the 1950s in spite of the higher prices charged for them.

(4) ... は語や節の省略を意味する．終止符で終わるときは .... となる．

R

radiation
放射に用いる物理量の記号と SI 単位については付表 9 を見よ．

radical
(1) **遊離基**の化学式において，不対電子を表すのに中黒 centered dot が用いられる．ただし，右上に dot をつけてもよい．

| ·H | ·CH$_3$ | ·SH | ·C$_6$H$_5$ |
| H· | CH$_3$· | HS· | C$_6$H$_5$· |

(2) 遊離基が電荷を帯びるとき，すなわち，**陽イオンラジカル**，**陰イオンラジカル**の場合には右上に dot をつけ，続いて電荷を記す．

R$^{·+}$ R$^{·-}$ C$_3$H$_6$$^{·+}$

ただし，質量分析では・と正負の記号の順序が逆となる．

rather
(1) 強意語として，形容詞，副詞の前におかれる副詞である．
▶ Imperfections normally exist in rather high concentrations in all real materials and may be classified as point defects, dislocations, or line defects, and plane defects.

(2) than を伴って，**むしろ**の意味を表す．
▶ There are also marked differences between mercury and cadmium, but often in degree rather than in kind.

ratio
比は数字と：または / を用いて表す．：または / の前後は詰める．

| a ratio of 1:10 | a ratio of 1/10 |
| a 1:2 (v/v) mixture | a 1/2 (v/v) mixture |

re-
(1) **また，再び，新たに**の意味の接頭辞で，一般にハイフンは用いない．

| rearrangement | recombination |
| recrystallization | regenerate |

(2) 例外は，prefix (2) に記した場合．

re-emit   re-emphasize   re-collect

reason
(1) reason is that の方が reason is because より好ましい表現とされる．
▶ The <u>reason</u> for the factor 1/2 <u>is that</u> otherwise we would count each collision twice.

(2) reason that の方が reason why より好ましい表現とされる．
▶ The <u>reason that</u> enthalpies of adsorption must be negative is that the adsorption process inevitably involves a decrease in entropy.

reduce
(1) **簡単な形にする**の意味の他動詞である．
▶ The complicated concentration dependence of this equation cannot <u>be reduced</u> to a simpler expression.

(2) **還元する**の意味の他動詞である．
▶ To <u>reduce</u> a carbonyl group that is conjugated with a carbon–carbon double bond without <u>reducing</u> the carbon–carbon double bond, too, requires a regioselective <u>reducing</u> agent.
▶ Aldehydes and ketones can <u>be reduced</u> to alcohols by molecular hydrogen in the presence of metal catalysts such as platinum, palladium, or nickel.

refer to
…**のことを指す**の意味の自動詞，**照会させる**の意味の他動詞である．
▶ Just as $[uvw]$ <u>refers to</u> a direction which is unspecified, so (hkl) <u>refers to</u> a plane which is unspecified.
▶ A pair of electrons occupying a nonbonding AO commonly <u>is referred to</u> as lone pair.

references and notes
引用文献と注である．注がない場合，速報誌（Chem. Lett.）では and notes を省く〔☞ citation in text〕．

(1) 論文誌名を国際的慣習に従って略記するときには終止符を使用する．さらに，references and notes で使用する次の略語にも終止符を用いる．

    Chapter (Chap.)　　　edition と edited (ed.)
    page (p.)　　　　　　pages (pp.)　　　volume (Vol.)

(2) 複数の著者名はすべて記入する．コンマで区切るだけで and は使用しない．

(3)　論文誌の名称が途中で変更されているときは，発表当時の誌名で表示する．例えば

　　　Ber.　　　　　　　　　　　→ *Chem. Ber.* (1947–)
　　　J. Chem. Soc., Chem. Commun. → *Chem. Commun.* (1996–)
　　　Q. Rev. Chem. Soc.　　　　→ *Chem. Soc. Rev.* (1972–)

(4)　ibid., idem は使用せず，すべての論文誌名を略記する．

(5)　ハイフンを用いた雑誌の略記には，全角ダッシュを用いる．

　　　Chem.——Eur. J.　　　　　*Catal. Rev.——Sci. Eng.*
　　　J.——Am. Water Works Assoc.　*J.——Assoc. Off. Anal. Chem.*

(6)　文献番号は，1, 2, 3, …，とし，かっこも終止符も使用しない．

(7)　引用文献は「ローマン体の著者名，イタリック体の論文誌名の略記，ゴシック体の発行年，巻数表示があればイタリック体の巻数，ローマン体の開始ページ数」の順に記する．略記された論文誌名と発行年の間にはコンマは使用しない．合併号の場合，例えば33巻と34巻ならば，*33–34* と記する．

1　N. P. Balsara, L. J. Fetters, N. Hadjichristdis, D. J. Lohse, C. C. Han, W. W. Graessley, R. Krishnamoorti, *Bull. Chem. Soc. Jpn.* **1999**, *32*, 6137.

2　S. Mori, K. Uchiyama, Y. Hayashi, K. Narasaka, E. Nakamura, *Chem. Lett.* **1998**, 111.

(8)　同一文献番号で複数の論文を引用するときは，セミコロンで区切る．あるいは，終止符で区切って，それぞれの論文の前に a), b), c) …を入れる．

3　F. Minisci, *Synthesis*, **1973**, 1; G. A. Olah, T. D. Ernst, *J. Org. Chem.* **1989**, *54*, 1203; H. Takeuchi, T. Adachi, H. Nishiguchi, *J. Chem. Soc., Chem. Commun.* **1991**, 1524; S. Seko, N. Kawamura, *J. Org. Chem.* **1996**, *62*, 442.

4　a) H. Tsutsui, Y. Hayashi, K. Narasaka, *Chem. Lett.* **1997**, 317. b) K. Uchiyama, M. Yoshida, Y. Hayashi, K. Narasaka, *Chem. Lett.* **2004**, *33*, 607. c) M. Kitamura, S. Chiba, Narasaka, *Bull. Chem. Soc. Jpn.*, in press.

(9)　単行本（書籍）を引用するときは，イタリック体で書名を記し，前後を引用符で区切ることはしない．

5　G. A. Olah, G. K. S. Prakash, J. Sommer, *Superacids*, John Wiley & Sons, New York, **1985**, p. 243.

6　J. K. Stille in *The Chemistry of the Metal-Carbon Bond*, ed. by F. R. Hartley,

Wiley, Chichester, **1985**, Chap. 6.
7 G. D. Wignall in *Encyclopedia of Polymer Science and Engineering*, 2nd ed., ed. by H. F. Mark, N. M. Bikales, Wiley-Interscience, New York, **1999**, Vol. 10, Chap. 6, pp.112−150.

(10) 学位論文の引用例
8 K. W. Mess, Ph. D. Thesis, University of Leiden, Leiden, The Netherland, **1969**.

(11) 政府刊行物の引用例
9 *Selected Values of Chemical Thermodynamic Properties*; NBS Technical Note 270-8, U. S. Department of Commerce, Washington, DC, **1981**.

(12) 特許明細書の引用には，Chemical Abstracts reference を必ず添える．
10 H. F. Lockwood, U. S. Patent 3759835, **1965**; *Chem. Abstr.* **1970**, 73, 46241q.

(13) 学術講演（予稿集）の引用例
11 Presented at the 63rd Annual Meeting of the Chemical Society of Japan, Higashi-Osaka, March 23–27, **1992**, Abstr., No. 2171.

(14) 注 notes の例
12 The hydrogen isotope effect in this exchange reaction can be explained by conventional theory. See, for example, M. M. Kreevoy, *J. Chem. Educ.* **1964**, *41*, 636 and references cited therein.

and references cited therein は，引用文献のページ数に，コンマを用いずに続けて記載する．cited を省いた and references therein も可．

13 That ρ is not 1/3 was first pointed out by J. F. Nagel (private communication, 1973) on the basis of approximate calculations.
14 A fuller discussion of this notion and additional examples will be published in *Bull. Chem. Soc. Jpn.*, in press.
15 Crystallographic data for **4a** have been deposited with the Cambridge Crystallographic Data Centre as supplementary publication number CCDC-000000.

regardless of
…にかかわらずの意味である．

▶ Diamagnetism is a property possessed by all atoms regardless of what other type of magnetic behavior they may exhibit.

relative to
　…と比較して，…に比例しての意味で用いたとき，文法上，対等な関係にあるものを取り上げるよう注意する〔☞ and〕．
- ▶ The lifetime of the transition state is so short relative to the time scale of the majority of our investigation techniques that direct observations are normally not possible.

relatively
　厳密には関係を表し，**比較的**，**割合に**を意味する副詞である．**いくぶん**，**やや**を意味する rather, somewhat の代わりに用いることは避ける．
- ▶ There are certain resemblances among metal ions that can be discussed in terms of oxidation state but which are relatively independent of electron configuration.

respect to, with
　…に関して，についてはの意味である．
- ▶ If proton is accepted as a unique particle, even with respect to a nonaqueous solution, an acid may be defined as a substance that can give up a proton.
- ▶ The viscous effect is just the difficulty of moving one part of a fluid with respect to another part.

respectively
　それぞれ，**おのおの**，**めいめいに**を意味する副詞で，複数の語の間の対応を明示する目的で用いられる．通常は文末にコンマに続いておかれる．次の例文にも見るように，必ずしも必要としない．多用は避けるのが望ましい．
- ▶ In the crystal of **1**, the intramolecular $O(1)\cdots O(9)$ and $O(4)\cdots O(10)$ distances are 2.566 and 2.576 Å, respectively; the corresponding $H\cdots O$ contacts are 1.67 and 1.81 Å, and the $O-H\cdots O$ angles are 146 and 145°. In **2**, the intramolecular $O(1)\cdots O(9)$ and $O(4)\cdots O(10)$ distances are 2.545 and 2.546 Å, while the corresponding $H\cdots O$ contacts are 1.88 and 1.79 Å, with $O-H\cdots O$ angles of 143 and 150°.

restrictive clause, nonrestrictive clause
　関係代名詞（who, which, that, what），**関係形容詞**（whose, which, what），**関係副詞**（when, where, why, how）を用いたとき，これに続く節が先行詞の意味を制限する場合と制限しない場合がある．

(1) 意味を伝えるのに欠くことができない**制限的**な場合には，節の前にコンマを用いない．which ではなく that を用いるのが望ましいとされるが，前置詞を組み合わせる場合には前者を用いる〔☞ which (2)〕．

▶ It is known that an ion has a negative contribution of entropy which is due to the restriction of degrees of freedom of solvent molecules in the vicinity of the ion and that this effect is greater the greater the charge of the ion.

▶ The primary significance of bond energies lies in the calculation of the enthalpy of a reaction involving a compound for which no enthalpy data are available.

(2) 補足的な説明であって，なくても文の意味は失われない**非制限的**な場合には，節の前後をコンマで区切る．この場合に用いられる関係詞は who, which, when, where の四語に限られる．

▶ The words that are used in describing nature, which is itself complex, may not be capable of precise definition.

result
結果として生じる，起因する，帰着する，終わるの意味の自動詞である．

(1) result from は**起因する**の意味で用いる．

▶ The driving force of a chemical reaction results from a tendency for the system to approach equilibrium.

(2) result in は結果的に**終わる**の意味で用いる．

▶ Autoionization of sulfuric acid results in the formation of the hydrogen sulfate ion and a solvated proton.

roman type
論文の主体は**ローマン体**で印刷されるが，特に注意すべきものを次にまとめる〔☞ boldface type, italic type, Latin term〕．

(1) 普通に用いられるラテン語はローマン体とする．

ab initio	ad hoc	a priori
in situ	in vitro	in vivo
status quo	vice versa	

(2) ラテン語を略記するときもローマン体とする．

a.m. (ante meridiem)	ca. (circa)
cf. (confer)	e.g. (exempli gratia)

et al.（et aliti）　　　　　etc.（et cetera）
　　　i.e.（id est）　　　　　　　vs.（versus）
（3）　SI 単位および非 SI 単位記号にはローマン体を用いる．
　　　K　　　　　　mmHg　　　　Pa
（4）　一般には，cis, trans, syn, erythro などにはローマン体を用いる．
ただし，化学物質名の一部であるときには，イタリック体とする．
　　　cis–trans isomerism　　　syn form　　　erythro diol
（5）　名詞や形容詞の前に，ハイフンを用いてつけた元素記号はローマン体
とする．
　　　N-alkylation　　　S-substituted
（6）　π, e, i などの数学的定数，d, ∂ などの演算子はローマン体とする．
（7）　三角関数，指数関数，対数など，関数の記号はローマン体とする．

cos	cosh	cot	coth
div	exp	grad	lim
ln	log	max	min
sec	sech	sin	sinh
tan	tanh	tr	

（8）　数値を表す数字は，ローマン体とする．ただし，文頭の数値には，数字を用いず，数詞を綴る．
（9）　S_0, S_1, T など系の量子状態を表す文字記号は大文字のローマン体，s, p, d など 1 個の粒子の量子状態を表す文字記号は小文字のローマン体とする．

S

same

(1) **同じ**，**同一の**を意味する形容詞で，比較級はない．または**同一のもの**を意味する代名詞である．通常，the をつけて用いる．

- In an isolated gaseous ion of a transition metal, the d orbitals are degenerate, i.e., they are all of the same energy.
- The alkyl groups in secondary and tertiary aliphatic amines can of course be the same or different; these amines are usually prepared by appropriate alkylation at a nitrogen atom.

(2) the same (...) as, the same ... that は**…と同じ**，**…と同様の**の意味を表す．

- An important limitation for MX (1:1) compounds is that the coordination number of the cation is the same as that of the anion.
- The problem of the determination of individual values of ionic entropies poses much the same difficulties as ionic enthalpies.

-scope

観察のための機器，**…鏡**の意味の接尾辞で，名詞を作る．

    oscilloscope   telescope

-scopic

見る，**観察する**の意味の接尾辞で，形容詞を作る．

    macroscopic   microscopic

-scopy

観察，**観察法**の意味の接尾辞で，不可算名詞を作る．

    microscopy   spectroscopy

seem

…らしい，**思われる**の意味の自動詞で，形容詞，不定詞，that などがこれに続く．

- There seems to have been no detailed microscopic observation of oriented single crystals of resorcinol as they underwent polymorphic transformation.
- It seems that rubidium differs markedly from potassium in only one respect:

the formation of metallic oxides.

semicolon

セミコロン；はコンマよりは終止符にやや近い切れ目で，その後は1字分あける．

(1) 接続詞のない節の間の区切りに用いる．

▶ The various isotherms of the Langmuir type are based on the simplest of assumption; all sites on the surface are assumed to be the same, and there are no interactions between adsorbed molecules.

▶ Phenol may be converted into a mixture of o- and p-nitrophenols by reaction with dilute nitric acid; the yield of p-nitrophenol is increased if a mixture of sodium nitrate and dilute sulfuric acid is employed.

(2) コンマを含む語，節，測定値の列記を，さらに区切るときに用いる．

▶ Common among physical methods are pressure measurements in gaseous reactions; dilatometry, or measurement of volume change; optical methods such as polarimetry, refractometry, colorimetry, and spectrophotometry; electrical methods such as conductivity, potentiometry, polarography, and mass spectrometry.

▶ Found: C, 47.69; H, 3.94; N, 20.98%. Calcd for $C_{21}H_{20}N_8O_9$: C, 47.75; H, 3.81; N, 21.24%.

▶ **Figure 1.** The variation of rate with concentration for various types of surface reactions: (○), simple unimolecular process; (□), a bimolecular reaction occurring by a Langmuir–Hinshelwood mechanism; (△), a bimolecular reaction occurring by a Langmuir–Rideal mechanism.

(3) 独立した節の間を結ぶ副詞ないし接続詞 hence, however, moreover, namely, nevertheless, nonetheless, that is, therefore, thus などの前にはセミコロン，後にはコンマを用いる．

▶ The molecular complex as a whole determines the color; that is, the light absorption causing color is not localized in one of the partners.

(4) 主節と従属節の間を結ぶ接続詞 after, although, as, as if, as soon as, because, before, if, like, since, though, till, unless, until, whereas, while などの前には，セミコロンを用いてはいけない．

series of, a
一続き，連続の意味の集合名詞で，一つの集団と見なす場合には動詞の単数形と，個々のものを強調する場合には動詞の複数形と組み合わせる．
- There is a series of analogous cyclic thiophosphoric acids with the formula $(HS_2P)_n$ that may be prepared by the oxidation of red or white phosphorus with polysulfides under a variety of conditions.
- A series of closely related structures are found for complex halides with the perovskite structure.

set of, a
一組，一式の意味の集合名詞で，動詞の単数形と組み合わせる．
- When a set of hybrid orbitals is constructed by a linear combination of atomic orbitals, the energy of the resulting hybrids is a weighted average of the energies of the participating atomic orbitals.

similar, like
(1) **同様の，同種の**の意味の形容詞である．
- Solutions of organic substances are almost invariably made in organic solvents, simply because here again the attractive forces between two different types of molecule are likely to be large if the molecules are similar.
- The relationship between bulk silicon and the silicon atom is similar to the relationship between organic polymers and their constituent monomers.

(2) similar to と同じく，**…のような**の意味をもつ like は形容詞のほか，副詞にも用いられるが，similar は形容詞だけである．したがって，副詞を必要とするときには，like もしくは similarly to を用いる．
- Like many other facts of nature, the insolubility of hydrocarbon liquids in water is easy to observe and describe but unexpectedly difficult to explain.

since
…以来の意味ではなく，次の例文のように，**だから**の意味で多く用いられる．ただし，ACS は since を前者の意味で用い，後者の意味の場合は because を用いることが望ましいとする．
- Since any group of atoms that gives up a proton is called an acid, acids may be cations, anions, or neutral molecules.
- The methyl nitrite self-decomposition flame is suitable for study at

atmospheric pressure <u>since</u> it has a low burning velocity and a wide reaction zone.

singular and plural

主語が**単数形**であれば動詞も単数形，主語が**複数形**であれば動詞も複数形と，数の一致を要することは，誰でも承知していることであるが，現実には誤りやすい．

(1) 主語と動詞が近接しているとは限らないことに注意する．

(2) 主語として，複数個の単数形が and で連結されているときには，動詞は複数形を用いる．これには例外がある〔☞ and (2)〕．

(3) 主語として複数の語が or で連結されているときには，動詞の数は or に最も近い語の数に一致させる．

(4) 集合名詞は，一つの集団と見なすときには単数として扱い，集団を構成する個々を意味するときには複数名詞として扱う〔☞ noun (2), majority, series, variety〕．

(5) data は単数形としても，複数形としても扱われる〔☞ data〕．

(6) 測定単位は集合名詞として扱われるから，動詞は単数形を用いる．

▶ To a hot solution of 104 mg of TCNQ in 15 mL of tetrahydrofuran was added 100 mg of anthracene.

(7) 科学用語の ics で終わる名詞は単数形である．

dynamics　　　　kinetics　　　　mathematics
mechanics　　　　physics　　　　thermodynamics

(8) anybody, anyone, each, either, everybody, every one, everyone, neither, no one, somebody, someone を主語とするときは，動詞は単数形を用いる．

(9) all, any, most, none, some を主語とするときは，動詞に単数形を用いるか，複数形を用いるかは文脈による．most, none, some については，それぞれの項の説明を見よ．

(10) both, few, many, several を主語とするときには，動詞は複数形を用いる．

(11) 主語が，あるものの分数あるいは一部であるとき，そのあるものが単数形か複数形かで，主語の数は定まる．

　　one-third of the precipitate　　one-third of the electrons

slash

スラッシュ/は次の場合に用いる．その前後は詰める．

(1) 二分ダッシュと同じく，混合物の成分の間を区切るのに用いる．

glycerin/water　または　glycerin–water

a metal/ligand reaction mixture

　　または　a metal–ligand reaction mixture

a methane/oxygen/argon matrix

　　または　a methane–oxygen–argon matrix

(2) コロンと同じく，比を表すのに用いる．

1/1　または　1:1　　　　　1/50/450　または　1:50:450

a/b　または　$a:b$

(3) 単位の表現で，per を表すのに使用する．

dm^3/mol　または　$dm^3\,mol^{-1}$　　　$J/(mol\,K)$　または　$J\,mol^{-1}\,K^{-1}$

(4) 表の各列に示される数値は物理量/単位であるから，物理量，単位それぞれの記号を用い，次のように記して列の冒頭におく．

T/K　　　$E/kJ\,mol^{-1}$

(5) 上付きまたは下付きの分数に用いる．

$x^{1/3}$　　　　$(x+y)^{1/2}$　　　　$t_{1/2}$

so

(1) **それほど，非常に**の意味で，形容詞の程度を強調するのに用いる．

▶ Since a transition state is <u>so</u> unstable, and since it has such an extremely short lifetime, there is no way to determine its precise structure.

▶ We rightly expect lanthanum to be an electropositive metal, less <u>so</u> than barium but more <u>so</u> than yttrium.

(2) so...that... は，程度，結果を表して**…なほど…で，非常に…なので**の意味である．

▶ The plasma frequency depends on the free-electron concentration, and it is <u>so</u> because this is slow for solids such as indium tin oxide <u>that</u> they have a plasma frequency in the IR part of the spectrum and so are transparent in the visible.

(3) and so は接続詞的な用法で，**それゆえに，だから**の意味である．

▶ Tantalum is even more resistant to corrosion than is niobium <u>and so</u> is used

in chemical plants and for surgical implants.
▶ Pyridine was added to remove the HCl <u>and so</u> prevent the reverse reaction.

some
(1) 複数名詞の前に some をおいたときには，誰か，何か，いくらかは特定されない．それが主語であれば，動詞は複数形を用いる．
▶ An interesting phenomenon found in <u>some</u> crystals is the tendency to be pyro- and piezoelectric.

(2) 不可算名詞の前に some をおいて，特定されない分量を表す．この場合の動詞は単数形を用いる．
▶ <u>Some</u> evidence for p_π–d_π bonding in silicon compounds is provided by the shapes of the molecules $N(SiH_3)_3$ (planar) and H_3SiNCO and H_3SiNCS (linear).

(3) some は代名詞の単数形とも複数形ともみなすことができる．

somewhat
やや，多少の意味の副詞で，rather と同様に用いられる．
▶ The theory of the glass electrode is <u>somewhat</u> complicated, but when the bulb is inserted into an acid solution, it behaves like a hydrogen electrode.

space and time
空間と時間に用いる物理量の記号と SI 単位については付表 10 を見よ．

specific rotation
比旋光度 (specific optical rotatory power) α は，右下の測定に用いたナトリウム D 線，右上の測定温度 (°C)，かっこ内の濃度 c (単位は g/100 mL)，使用した溶媒 dioxane を含めて，次のように記す．

$[\alpha]_D^{25} +26.4° (c\ 1.00, dioxane)$

spite
…にもかかわらずの意味で，in spite of ... や despite の形で用いられる．
▶ <u>In spite of</u> the apparent success of the dielectric constant in correlating electrolyte behavior, we should not get the impression that solution properties can be understood in terms of this parameter alone.

sub
(1) 下位，副，亜の意味の接頭辞で，一般にハイフンを用いず一語とする．
 subgroup   sublevel   suboxide

(2) 例外は b が重なる場合
sub-bandwidth

subscript

下付き文字には，イタリック体とローマン体がある．これをつけられる文字との間は詰める．印字が不明確な場合は，下付きであることを記載する．

(1) 物理量 P, V, T や数 x, n を表すときのみ，イタリック体とする．

heat capacity at constant pressure　C_P
heat capacity at constant volume　C_V
isothermal compressibility　κ_T
isoentropic compressibility　κ_S
$KAl_xCr_{1-x}(SO_4)_2 \cdot 12H_2O$　　M_xWO_3　　M_nO_{3n-1}

(2) ローマン体の下付き文字として，熱力学で用いられるものは

ads（adsorption）　　　at（atomization）
c（combustion）　　　 dil（dilution）
dpl（displacement）　　f（formation）
fus（fusion）　　　　　imm（immersion）
mix（mixing）　　　　 r（reaction）
sol（solution）　　　　sub（sublimation）
trs（transition）　　　 vap（vaporization）

(3) $C^h_{P,b}$ と $C^h_{P,hs}$ の上付き文字，下付き文字に関する記述を次に示す．

▶ The superscript h identifies a hydrogen-bond-breaking contribution, and the subscript b refers to bonds in bulk water as opposed to those in the hydration shell, where the corresponding variables will be labeled with a subscript hs.

such as

…のようなを意味し，同種類のものを例示する場合に用いる．

(1) **制限的**な用法では，such as の前にコンマを用いない．

▶ In structures such as ice, each oxygen is attached tetrahedrally to four hydrogens, because this is required for the maximum number of hydrogen bonds to be formed.

(2) such as の後には，文法上，対等な関係にあるものを並べる．それらを and でつなぐか，or でつなぐかを検討する．

- ▶ Natural fibers such as cotton and human hair can feel very rough after washing.
- ▶ A continuum light source such as a xenon arc or a tungsten-halogen lamp illuminates the entrance slit of a high-luminosity grating spectrometer.

(3) 補足説明をする**非制限的**な用法では，コンマで区切る．
- ▶ A theory, such as the atomic theory, usually involves some idea about the nature of some part of the universe, whereas a law may represent a summarizing statement about observed experimental facts.

さらに，次のように such as を使用しない例もある〔☞ e.g.〕．
- ▶ Thermodynamics is a reliable guide in industrial chemistry, in plasma physics, in space technology, and in nuclear engineering, to name but a few applications.

such ... as ...
…のような…の意味である．
- ▶ Pauling proposed an electroneutrality principle stating that electrons are distributed in a molecule in such a way as to make the residual charge on each atom zero or very nearly zero.

suffix
接尾辞（連結形と呼ばれるものを含む）には，一般にハイフンを用いない．

able	ability	ance	escent
fold	ful	fy	gen
hedral	ium	ize	less
like	lysis	ment	mer
ness	ode	oid	ous
ship	tactic	wide	wise

(a) like には例外が多い〔☞ like〕．
(b) fold は 10 倍以上の場合は数値を用い，ハイフンで連結する．

superscript
上付き文字ないし記号をいう．これをつけられる文字との間は詰める．以下に，熱力学において推奨される上付き文字ないし記号を示す．

　　′ (apparent)　　　　　　E (excess quantity)
　　id (ideal)　　　　　　　∞ (infinite dilution)

* (pure substance)　　　°, ~ (standard state)

syllabication

語を**音節**に分けること，すなわち**分節法**のことをいう．

(1) ice, iron, mole, pure, strength, white, zinc などの単音節語は分割できない．

(2) 行末で語を分割して次行に送るときには，必ず音節の切れ目で行う．おおよそを Webster 系統の辞典によって示す．

 (a) 音節は発音する母音（二重母音を含む）一つを含む．
   ac・ti・nide   ox・i・dize   prep・a・ra・tion
 (b) 接頭辞，接尾辞は語幹から分けられる．
   co・po・ly・mer・ize   pro・to・ac・tin・i・um
 (c) 子音が重なる場合，分けられることが多い．
   am・mo・nia   ben・zene   cop・per

分節法は複雑である．辞典によっては，ハイフンを使用するならば，ここという音節とハイフンの使用が望ましくない音節を区別している．

関連した語でも，品詞によって音節の区切りが異なることがある．

  at・om   a・tom・ic   at・om・i・za・tion
  cat・a・lyst   cat・a・lyt・ic   ca・tal・y・sis
  mol・e・cule   mo・lec・u・lar

米国と英国で発音が異なるため，音節が異なる場合もある．

  mea・sure　(Webster)   meas・ure　(Collins)
  meth・ane　(同上)   me・thane　(同上)

印刷の際に行末が移動すると，原稿の行末にあるハイフンは，その使用理由を判断することなく残され，新しい行末にハイフンが使用される．校正にあたっては，著者はこれらの事情に注意を払うことが必要である．

(3) ハイフンの箇所でも，行末や行頭に 1 字だけとなることは避ける．

(4) 既にハイフンをもっている語は，そのハイフンの箇所で分割する．

  co-worker   half-life   self-consistent

(5) 略語は分割してはならない．

(6) 数字は分割してはならない．例えば，2004 を 20- と 04 に分けてはいけない．ただし，あまり桁数の多い数字は 3 桁ごとのコンマの次で切ることが許される．例えば，1,000,000,000 を 1,000,- と 000,000 に分ける．

(7) 辞典には固有名詞の音節も示されているが，分割は望ましくない．

| Ein·stein | Ev·ans | Mac·lau·rin |
| Ost·wald | Rob·in·son | Wil·liams |

symbols

数学の記号の使用にあたっては equation を見よ．

$\propto$, $\sim$	proportional to
∞	infinity
$\neq$	not equal to
$\approx$	approximately equal to （本文中の about は ca. と略記）
$\rightarrow$	approaches
$\parallel$	parallel to
$\perp$	perpendicular to
Σ	summation
Π	product
$\int$	integral
$<$	less than
$\leq$	less than or equal to
$>$	greater than
$\geq$	greater than or equal to
$\exp x$	exponential of x
$\ln x$	natural logarithm of x
$\log x$	logarithm to the base 10 of x
$\lvert a \rvert$	absolute magnitude of a
$\langle a \rangle$, $\bar{a}$	mean value of a
$f(x)$	function of x
∂x	partial differential of x
$\mathrm{d}x$	total differential of x

symmetry elements and point group

Schönflies の記号を用いて，対称要素に回転軸，開映軸，対称心，対称面を扱うと，対称面の σ を除き，記号はイタリック体のアルファベットとローマン体の数字からなる．

回転軸　C_1, C_2, C_3, C_4, C_6

開映軸　S_4

対称心　i

これらの組合せである 32 種の**点群**（**結晶群**）の記号についても同様である．

C_1, C_i, C_2, C_s, C_{2h}, D_2, C_{2v}, D_{2h}, C_4, S_4, C_{4h},

D_4, C_{4v}, D_{2d}, D_{4h}, C_3, C_{3i}, D_3, D_{3v}, D_{3d}, C_6, C_{3h},

C_{6h}, D_6, C_{6v}, D_{3h}, D_{6h}, T, T_h, O, T_d, O_h

synthesis

（1）　**合成**とは元素または簡単な化合物から複雑な化合物を作ることをいう．

▶ The synthesis of ethyl 1-naphthoate illustrates the preparation of a carboxylic ester by a variant of the Grignard carboxylation route to a carboxylic acid. The arylmagnesium bromide is first prepared and added to an excess of diethyl carbonate, conditions which minimize the possibility of further reaction of the Grignard reagent with the ester initially produced to form a tertiary alcohol.

（2）　**製法**を意味する名詞 preparation は，一つの物質の新しい状態，例えばコロイド粒子やナノ粒子を作ることにも用いられる．

▶ Hydrolysis is the usual method of preparing colloidal iron(III) hydroxide. However, the preparation fails unless an appreciable proportion of iron(III) salt, such as oxychloride, is retained as stabilizing agent.

（3）　粒子の配列などには，**組立て**，**製造**を意味する名詞 fabrication を用いる．

T

table
(1) **表**の題目 caption は簡潔に内容を伝えるものとし，文章の形にはしない．本文を参照せずに理解でき，本文にない情報は含まないものとする．題目は大文字ではじめ，終止符はつけない〔☞ caption〕．

　　Table 1. Crystal data for compounds **3** and **6**

(2) 表の各列の冒頭は，記号/単位であることが望ましい．物理量ではない記号を表で使用するときには，表の脚注で定義する．

(3) 表でのみ使用される略語は表の脚注で定義する．なお，概要，本文で用いる略語は，それぞれ概要，本文中で定義する〔☞ abbreviation (4)〕．

(4) 各列の見出しは，左端を上下に揃え，数値は小数点を上下に揃える．

(5) 説明を加えたい場合には，a, b, c ... を右肩につけて，その表の脚注とする．脚注は大文字ではじめ，終止符で終わる．

term
(1) 数式における**項**を意味する．

▶ The semiempirical approach requires only that a suitable <u>term</u> representing these attractions be inserted.

(2) in terms of は…の見地から，…の点から，…に関しての意味で用いられる．

▶ The principal difficulty with the bond-moment analysis is that the electron distribution that results in the molecular dipole moment cannot always be treated <u>in terms of</u> separate, noninteracting components.

▶ Thus, the production of bromoform is nicely rationalized <u>in terms of</u> inductive, resonance, and steric effects, the three most important factors chemists use to explain relative reactivities of organic compounds.

than
比較級の形容詞または副詞の後に使用する接続詞である．比較は文法上，対等な関係にある語について行う．この条件を満たすに必要な語を省略しないように注意せよ．

▶ The specific conductivities of these true solid electrolytes are several orders

- of magnitude higher than those that have their origin in point defects.
- ▶ The calixarenes are characterized by much higher melting points and lower solubilities in common organic solvents than their acyclic counterparts.

that

指示代名詞（複数形は those），**関係代名詞**〔☞ restrictive clause, non-restrictive clause〕，あるいは**接続詞**として用いられる．

- ▶ An example of a reaction that is extremely slow at room temperature is that between hydrogen and oxygen.
- ▶ The characteristic of the solid state is that the substance can maintain itself in a definite shape that is little affected by changes in temperature and pressure.
- ▶ Of all condensed-phase phenomena those of pure crystalline materials are most easily treated.

that is

すなわちまたは**例えば**の意味である．これに節が続く場合は，前にセミコロン，後にコンマを用いるが，その他の場合（例文が示すように，様々な場合がある）には，前後をコンマで区切る．

- ▶ Accordingly, substances with molecules of high symmetry crystallize more readily than those of low symmetry; that is, they have higher melting point.
- ▶ Frequently, it is desirable to estimate the enthalpy of a chemical reaction involving a hitherto unsynthesized compound, that is, a substance for which no enthalpy data are available.
- ▶ The polarizability of the anion will be related to its softness, that is, to the deformability of its electron cloud.
- ▶ The early studies on the structure of crystals were of necessity morphological, that is, based on the external appearance of the crystal.

that is の意味のラテン語 id est を略記した i.e. も，同様に使用される〔☞ i.e.〕．

the

(1) 既に言及したものや人には**定冠詞** the をつける．

(2) 前置詞句や関係詞節によって特定されたものや人には the をつける．その例は次の (3), (4) を見よ．

(3) **可算名詞**（普通名詞，集合名詞）の場合，すべてに通じる一般的なことを述べるときには，the をつけた単数形を主語に用いる．

▶ The H–H bond is the strongest known two-electron bond formed between identical atoms.

▶ The melting temperature is the temperature at which solid and liquid phases of a substance are in equilibrium at a given pressure.

あるいは，the をつけない複数形を用いる．

▶ Rotational transitions of molecules take place in the microwave region of the electromagnetic spectrum.

▶ Alkenes are readily converted by chlorine or bromine into saturated compounds that contain two atoms of halogen attached to adjacent carbons; iodine generally fails to react.

したがって，spectra では著者が測定したスペクトルを意味しない．特定するには，the spectra を用いる．

(4) **不可算名詞**（物質名詞，抽象名詞）の場合，一般的なことを述べるときには，the はつけない．

▶ Organic chemistry is the chemistry of the compounds of carbon.

▶ Photoelectron spectroscopy is the examination of the energy levels of molecules by determining the kinetic energies of electrons ejected by the absorption of high-frequency monochromatic radiation.

(5) the comparative adjective..., the comparative adjective は，…であればあるほど，それだけ余計に…するの意味で使用する．この構文は…すればするほどの節を前にする．簡潔であること，二つの部分が緊密であることが肝要である．それだけ余計に…するの節では，主語と動詞を倒置してもよい．ここに示す例文のように，動詞をまったく伴わないこともある．

▶ There is a useful rule in organic chemistry which says that the milder a reagent, the more selective it is.

▶ The lower the energy required for ionization of the molecule, and the more stable the molecular ion, the more intense will be the peak in the mass spectrum.

▶ Polycyclic aromatic hydrocarbons with both angular and linear types of ring fusion show absorption curves of a similar profile to that of benzene but with

the absorption maxima shifted to longer wavelengths; <u>the greater</u> the number rings <u>the more pronounced</u> the shift.
- ▶ The overall extent of thermal decomposition in sodium chlorate monocrystals depends very much on the sample perfection, <u>the lower</u> value being typical of <u>the more perfect</u> specimens.
- ▶ <u>The greater</u> the number of physical properties measured, <u>the stronger</u> the evidence.

therefore
それゆえ，したがってを意味する副詞である．
(1) 導入の語として文頭で使用した場合には，次にコンマを要する．
- ▶ <u>Therefore</u>, atoms are not likely to combine in such a bimolecular process but must do so in a termolecular collision where the third molecule removes energy in the form of translation or vibration.

(2) 二つの節を結ぶときは，however と同様に，前をセミコロンで，後をコンマで区切る．

(3) その他の場合，前後ともコンマで区切るが正しいとされている．
- ▶ In spite of these advances in chemistry, the details of the atom and, <u>therefore</u>, of many aspects of the molecule remained a complete mystery.

thermodynamics
熱力学に用いる物理量の記号と SI 単位については付表 11 を見よ．

this, that
これらの代名詞は何を指すかが明確なときにのみ用いる．不明確なときには名詞で記載する．
- ▶ The photochemistry of transition-metal carbonyl complexes has been the focus of many investigations. <u>This</u> is due to the central role that metal carbonyl complexes play in various reactions.

後者の主語である代名詞 this が何を指すか不明確である．したがって，this interest と改める．

though
…だけれども，にもかかわらずを意味する接続詞であって，however とは機能が異なる．
- ▶ The radius ratio is a useful, <u>though</u> imperfect, tool in our arsenal for

predicting and understanding the behavior of ionic compounds.

through

(1) **…を通って，…を貫いて**の意味の前置詞として用いる．

- ▶ Potassium–graphite was prepared by the method of Fredenhagen and Suck, which ensures that contact between the potassium and the graphite occurs only through the vapor phase.
- ▶ The inductive effect is transmitted along chemical bonds, while the field effect operates through space or, in solutions, through the solvent or the low-dielectric cavity provided by organic solutes.
- ▶ The oxidation states 0 and III through to VII are known for plutonium, so it is not surprising that numerous compounds exist.

(2) 手段，媒体を表す by means of や by の意味で用いることは避ける．したがって，次の例は望ましいものではない．

- ▶ Copper(II) chloride thus diffuses and allows the cell to decay through direct reaction of the electrode materials.
- ▶ The condensation of polyhedra through square faces is not such a common aggregation process.

thus

このように，したがっての意味の副詞である．導入の語として文頭で使用した場合には，次にコンマを要する．

- ▶ Thus, whereas stable neutral sulfur iodides and selenium iodides either are not known or can only be prepared at low temperatures, sulfur–iodine and selenium–iodine cations have been shown to be unexpectedly numerous.

till, until

ともに**…まで**の意味の接続詞であるが，後者の方が固い文体となる．

title

表題は明確かつ簡潔で，文法的に正確であることを要する．そのためには，(1)ないし(5)を念頭に入れて表題を選ぶ．表題の印刷形式は，論文誌によって様々である．以下に記す速報誌（Chem. Lett.）の形式は，欧文誌や米国化学会の発行誌のものに近く，比較的複雑なものに属する．ただし，著者が校正を行う前に，(6)以下の項目については，編集者によって整えられる．

(1) keywords を含める．

(2) 通常，表題のはじめに the は必要としない．終止符はつけない．
(3) 次の句の使用は避ける．
　　on the, a study of, research on, report on, regarding, use of
(4) 非定量的な，意味のない表現は用いない．
　　rapid　　　new
(5) すべての語を綴る．記号，式，略語は避ける．上付き，下付きを含む表現をしない．社名，商標などを使用しない．
(6) 名詞，代名詞，動詞，形容詞および副詞の頭字は大文字にする．
(7) 接続詞，冠詞および前置詞の頭字は小文字とする．
　(a) ACS は不定詞の前の前置詞 to の t は大文字とするとしているが，これには従わない．
　　Using Mechanistic and Computational Studies to Explain Ligand Effects in the Palladium-catalyzed Aerobic Oxidation of Alcohols
　(b) ACS は従属接続詞として用いた as は As とし，前置詞として用いたときは as とするとしているが，ともに小文字とする．
　　Hydrolysis of 2-Chloro-2-methylpropane as Determined by Conductivity Measurements
　　Alumina as a Catalyst
(8) 句動詞（動詞と副詞の結合）は両部分とも頭字を大文字とする．
　　Break Down　　　Line Up　　　Set Off
　　Slow Down　　　Stand By　　　Warm Up
(9) X-ray の r は大文字としない．しかし，α particle と β particle の p，γ ray の r は大文字にする．
(10) 化学物質名の一部をなすイタリック体の *cis*, *trans* は大文字にはしないで，物質名の頭字を大文字にする．
　　cis-Azobenzene　　　*trans*-Azobenzene
(11) mg は Milligram とする．mg を Mg とすると，10^{-3} g から 10^6 g へ意味が変化する．このように，小文字で記された単位の記号には，大文字に変えられないものがある．
(12) 複数の語からなる形容詞は，頭字のみを大文字とする．
　　Acid-catalyzed Reaction　　　　　　Low-temperature System
　　Non-hydrogen-bonding Molecules　　Two-dimensional System

End-on Bonding　　　In-plane Atoms　　　Out-of-plane Vibrations
（13）　単独の語として頭字を大文字とすべき成分からなる複合語の各成分には，大文字を用いる．
　　　Cross-Link　　　Half-Life　　　Quasi-Elastic
（14）　ハイフンを要する接尾辞は大文字にはしない．
　　　Adamantane-like　　　Morphine-like　　　Olefin-like
（15）　多量体の名称は，頭字であるpのみを大文字とする．
　　　Reactions of Poly(methyl methacrylate)
（16）　複雑な置換基をもち，かっこを使用して表された化学物質名では，頭字だけを大文字とする．
　　　Tetrakis(methyl isocyanide)iron Complexes
（17）　動植物，微生物の属と種の名はイタリック体を用い，属名の頭字は大文字とすることは，title においても変わらない．
　　　Active-site Nucleophile of *Bacillus circulans* Xylanase

too
　…**もまた**の意味の副詞で，強調したい語句の後におく．
- ▶ Here, too, there are some problems in the materials chemistry in that in order to prevent undue extension of the cable in use, high elastic modulus materials must be incorporated in order to take the strain.
- ▶ The particles in silica sols are spherical, and electron microscopic examination of gels shows that these too are based on roughly spherical particles, generally about 100 Å in diameter which are themselves composed of still smaller particles, about 15 Å in diameter.

trans
　ローマン体とイタリック体の使い分けと，ハイフンの使用については，cis, trans を見よ．

transport phenomenon
　輸送現象に用いる物理量の記号と SI 単位については付表 12 を見よ．

U

un

不，**反対**の意味を表す接頭辞である．

(1) 米式では，nが重なる場合も含めて，ハイフンを用いず，一語とする．

unbranched	uncertainty	unnamed
unnecessary	unnoticed	unoccupied
unpaired	unsaturated	unshared
unstable	unsymmetrical	unvulcanized

(2) un-ionize は unionize と区別するために，ハイフンを必要とする．

units

測定単位には，可能な限り **SI 単位**を使用する．

(1) 数字を伴わないとき，測定単位を記号で表さず，これを綴る．

a few centimeters	reciprocal seconds
several milligrams	volts per meter

文頭にあるか，表題の一部でない限り，頭字に大文字は使用しない．これらの場合に，per の代わりにスラッシュを用いてはいけない．

(2) 綴った単位の複数形には s をつける．ただし，次の単位の複数形は単数形と変わらない．他方，単位の記号は，単数，複数の両方を表すから，s をつけることはしない．

bar	hertz	lux	stokes	siemens

(3) 単位の中で，単語と記号を混用しないように注意する．例えば，mole fraction を mol fraction，newtons per meter を N per meter，a few milliliters を a few mL，100 F/m を 100 farad/m としてはいけない．

(4) 人名に由来する単位名は，その頭字も小文字で記する．かっこ内はその記号である．

ampere（A）	angstrom（Å）	becquerel（Bq）
coulomb（C）	curie（Ci）	dalton（Da）
debye（D）	fermi（f）	franklin（Fr）
gauss（G）	gilbert（Gi）	henry（H）
hertz（Hz）	joule（J）	kelvin（K）

newton (N)	oersted (Oe)	ohm (Ω)
pascal (Pa)	poise (P)	roentgen (R)
rydberg (Ry)	siemens (S)	sievert (Sv)
stokes (St)	tesla (T)	torr (Torr)
volt (V)	watt (W)	weber (Wb)

(5)　測定単位の記号を定義する必要はない．

(6)　記号はローマン体とし，記号の後に終止符をつけない．

(7)　**SI 基本単位**とその記号を記すと

ampere	A	electric current
candela	cd	luminous intensity
kelvin	K	thermodynamic temperature
kilogram	kg	mass
meter	m	length
mole	mol	amount of substance
second	s	time

次に，**SI 誘導単位**とその記号を記すと

coulomb	C	electric charge
farad	F	electric capacitance
henry	H	inductance
hertz	Hz	frequency
joule	J	energy, work, heat
newton	N	force
ohm	Ω	electric resistance
pascal	Pa	pressure, stress
siemens	S	electric conductance
tesla	T	magnetic flux density
volt	V	electric potential, electromotive force
watt	W	power, radiant flux
weber	Wb	magnetic flux

(8)　実験の記述以外では，day, week, month, year を略記しない．

(9)　次の単位を略記する記号はない．

bar	darcy	einstein

erg　　　　　　　　faraday　　　　langmuir

(10) 数字を伴うとき，単位は記号で表し，数字と記号の間はあける．

300 mL　　　　　　5 min　　　　　　3 Å

ただし，%，°，′，″は例外で，数字とこれらの間は詰める．

25%　　　　　　　45°30′　　　　　35′20″

(11) 数字と°Cの間はあけるが，°と C の間は詰める．

(12) 単位記号の間はあけるか，中黒・をおいて詰める．

200 m V　　　　　　200 m·V

(13) /単位記号は，/を省いて，単位記号に負の指数をつけてもよい．

4×10^{-9} /m　　　　　4×10^{-9} m^{-1}

同じ指数をもつ記号は一つのかっこに入れてもよい．例えば，joules per mole kelvin の記号としては，次のいずれかを用いる．

J (mol K)$^{-1}$　　　　　　J mol^{-1} K^{-1}

J/(mol K)　　　　　　　J·mol^{-1}·K^{-1}

(14) 表の各列の冒頭および図の軸には，物理量の記号/単位または著者が別に定義した記号/単位を記載する〔☞ table (2)〕．

(15) 数字と単位の組合せが形容詞として用いられる場合には，数値と単位の間をハイフンでつなぐ〔☞ hyphen (17)〕．

10-min exposure　　　　50-mL flask

100-W power　　　　　　25-mg sample

ただし，濃度と温度の単位は例外で，数字との組合せが形容詞となっても，ハイフンは用いない．

0.2 M KCl　　　　　　　3 N HNO$_3$

20 °C difference　　　　a 0.1 mol dm^{-3} solution

(16) 単位記号は表題や見出しの中では使用しないで，これを綴る．小文字の単位記号には，大文字に変えられないものがある〔☞ title (11)〕．

(17) 測定単位の表現の初めの部分が単位記号ではない場合は，次の単位記号との間にスラッシュを用いる．

counts/min　　　　　　lines/cm

(18) 測定単位の表現の後の部分が単位記号ではない場合は，前の単位記号とその語との間にスラッシュ，per のいずれを用いてもよい．

keV/channel　　　　　　keV per channel

(19) 次のような複雑な表現では，単位記号の前に per を用いず，スラッシュとする．

    30 μg of peptide/mL   20 mg of drug/kg of body weight

(20) 一つの単位には，10^n または 10^{-n} を意味する接頭辞は一つだけ用いる．接頭辞と単位との間は詰める．

    kilojoule or kJ   milligram or mg

10^n を意味する主な接頭辞とその記号を次に示す．

n	Prefix	Symbol
−15	femto	f
−12	pico	p
−9	nano	n
−6	micro	μ
−3	milli	m
−2	centi	c
−1	deci	d
1	deca	da
2	hecto	h
3	kilo	k
6	mega	M
9	giga	G
12	tera	T
15	peta	P

(21) 範囲を表すなど，複数個の値を記す場合，単位は最後につける．なお，範囲を表す二分ダッシュの前後は詰める．

    10–15 mg   5, 10, and 25 mL   60–90 K
    from 10 to 15 mg   between 25 to 35 min

(22) 単位は集合名詞として扱われるので，動詞は単数形を用いる．

▶ To 45 mL of 5 M aqueous hydrochloric acid in a 125-mL flask is added 4.5 g of potassium tetrachloroplatinate(II) (0.0108 mol).

(23) 数字と記号の組合せは単数形と見なすから，量に関する比較級の形容詞である less を用いる〔☞ less〕．

    less than 10 mg   less than 2 days

(24) 式に続いて，用いた変数の定義を行った場合，必要ならば，かっこの中に入れて単位を略記する．
▶ $$r = d/P_o$$
where r is the distance between particles (cm), d is the particle density (g/cm^3), and P_o is the partial pressure of oxygen (kPa).

unlike
…と違っての意味の前置詞である．これを用いた導入の句は，主文の主語を引き合いに出すものでなければならない．
- ▶ <u>Unlike</u> mechanical and electrical oscillators, chemical oscillators never pass through their final equilibrium configuration in the course of an oscillatory cycle.

これと比較すれば，次の文が正しくないことは明らかであろう．
- ▶ <u>Unlike</u> Sr and Zn, sources of Pb in general are unnatural, such as pewter containers, lead-glazed pottery, or Pb pipes.

usually
一般にを意味するが，**習慣的にいつも**の意味が強く，generally と相互に代用できるとは限らない．
- ▶ The purity of as-received organic materials is <u>usually</u> in the range 95–99 percent.

UV–visible spectroscopy
UV は ultraviolet の略で，UV–visible は UV–vis と略記する．**紫外-可視分光**の測定結果の記載例を示すと

UV (hexane) λ_{max}, nm (ε) : 250 (1080).

UV (C$_2$H$_5$OH) λ_{max}, (log ε) 220 (3.34), 252 (sh, 3.12), 298 (sh, 2.32), 432 nm (3.27)

ここで，λ_{max} は nm で表した吸収極大の波長，ε は吸光係数（モル吸収係数）で，単位を特に指定する必要はない．sh は shoulder の略記である．

V

variable
変数にはイタリック体を用いる〔☞ italic type (2)〕.
(1) 質量 m, 時間 t, 温度 T, 体積 V, モル分率 x など
(2) x の複数形は x values または x's
(3) 数式に続いて，これに用いた変数と定数の定義を行う．必要ならば，単位をかっこに入れて記載する．
▶ $$PV = nRT$$
where P is the pressure, V is the volume, n is the amount of substance, R is the gas constant, and T is the temperature.

variety
多様，いろいろの意味の名詞で，一つの集合と見なす場合には，動詞は単数形，個々のものを強調する場合には，動詞は複数形を用いる．前者の場合は the，後者の場合は a を伴うことが多いとされる．
▶ The variety shown by hydrogen is largely a result of the many ways it can use its 1s orbital or any molecular orbital derived from it.
▶ Conjugated dienes react with a variety of ethylene derivatives to give cyclohexenes, a reaction first investigated by Diels and Alder.

vector
大きさと方向をもつ**ベクトル**は，イタリックゴシック体の大文字で，その成分は角かっこに入れたイタリック体の小文字で表す．$V = [a, b]$

versus, vs.
…**対**，…**と対比して**の意味の前置詞である．
A plot of K^{-1} versus ε_c.
ACS は，図表の説明や本文中のかっこで囲まれた部分でのみ vs. (ACS とは異なり，終止符を使用) と略記し，本文では versus と記すとする．

via
…**を経て**の意味の前置詞として用いる〔☞ through〕.
▶ MoS_2 is converted, via the acidic oxide MoO_3, to ammonium molybdate, which can be reduced to the metal with hydrogen.

W

week
実験の記述以外では，day, month, year と同様に略記しない．

whatever
どんな…でもの意味の形容詞である．
▶ Whatever other methods are applied, the study of a thermal transformation can hardly be regarded as complete until X-ray techniques have been used.

whenever
…する時にはいつでもの意味の接続詞である．
▶ Whenever a reaction occurs with a change in the number or kind of ions present so that the electrical conductivity changes, measurement of the resistance offers a convenient and accurate means of following the course of the reaction.

where
（1）　関係副詞用法で…する所の意味を表す．制限的にも非制限的にも用いられる〔☞ restrictive clause, nonrestrictive clause〕．
▶ Short-range forces are dominant in adsorption phenomena of all kinds, so perhaps this is where the required knowledge is to be sought.

（2）　数式に用いた記号の説明をはじめるときに用いる．
▶ The Helmholtz energy, A, is defined as
$$A = U - TS$$
where U is the internal energy of a system, T is its temperature, and S is its entropy.

この説明中の where は in which で，its は the で置き換えてもよいが，is を等号＝で置き換えてはいけない〔☞ equation (2)〕．

whereas
ところがの意味の接続詞で，主節に対照的なことを述べる節を導く．
▶ Whereas the IR absorption spectra of weak complexes are effectively the sum of the spectra of the component molecules, in these strong interactions the IR spectra correspond closely to the sum of the spectra of the

corresponding ions, D^+ and A^-.

▶ Volume and heat capacity are typical examples of extensive variables, whereas temperature, pressure, viscosity, concentration, and molar heat capacity are examples of intensive variables.

wherever
…する所はどこでもの意味の接続詞である．

▶ Iron occurs in smaller quantities wherever rocks and the soils derived from them are brown, yellow, orange, or red.

whether
(1)　…かどうかの意味で，二者を選択する接続詞として用いられる．

▶ Questions have been raised as to whether bone mineral was analogous to the synthetic case or, in fact, consisted of an apatite phase alone with little, if any, amorphous phase present.

(2)　whether (...) or not は，肯定と否定との選択に用いられる．

▶ Some insight into whether or not real systems do behave in this way may be obtained by considering processes in which the kinetic barrier does not change with $\Delta G°$.

which
関係代名詞としては，制限的にも非制限的にも用いられる〔☞ restrictive clause, nonrestrictive clause〕．通常は，物を表す名詞を先行詞とするが，先行する全文，節，またはその一部を先行詞とすることもできる．which ではじまる非制限的な節の前，あるいは前後にはコンマをおく．

(1)　関係節の主語として用いる．

▶ The zincblende and wurtzite structures are of the type AB and differ in details which need not concern us.

(2)　above, at, below, by, for, from, in, into, of, on, through, to, with などの前置詞を伴う形でも用いられる．

▶ Kinetics is the study of the rates at which these bond-making and bond-breaking processes occur.

▶ Enantiotropic forms are those having a definite transition temperature, below which one form, and above which the other form, is stable, and at the transition temperature both forms are in equilibrium.

- Single-crystal X-ray studies are the usual way by which new solid-state products and structures are first characterized.
- The lack of a firm descriptive base in a great many systems, one from which one can rationally extrapolate a chemistry, means that synthesis is often an exploration in the real sense of the word.
- Coordination compounds in which the coordination number of the central ion is different from either four or six are much less common.
- Apparently, voids exist in the crystal lattice of the parent compound into which only molecules of certain dimensions can be inserted.
- When a reaction takes place on a surface on which there is a variation of activity, the overall rate is the sum of the rates on the various types of site.
- The rigid framework of iodides must be composed of polyhedra that share faces, thus forming a network of pathways through which the current carriers can move.
- A major uncertainty in studying reactions in the interstellar medium is the degree to which the heavy elements have been depleted in the gas by freezing onto the cold grains or by incorporation into the refractory grain material in the atmosphere of cool stars.
- Unfortunately, substances with which this hypothesis can be tested are rare, but some do exist, among them the acetylide ion.

while

科学論文では，…する間，…するうちの意味ではなく，**なのに比べて，一方では**の意味の接続詞に用いられることが多い．ただし，ACS は while は前者の意味で用い，後者の場合は although, whereas を用いるのが望ましいとする．

- The acid HOBr disproportionates in acidic solution, while the element itself disproportionates in alkaline solution.
- While electron-rich nucleophiles add to the carbonyl double bond, carbon–carbon double bonds of olefins are subject to attack only by electron-deficient species (electrophiles).

whilst

while と同義で，主として英国で用られる表現である．

- Whilst we can be sure from the laws of thermodynamics that an unfavorable energy change will not occur spontaneously, all we can say about a favorable one is that it might happen.

whole

全体を一つのまとまったものとして考える形容詞，または名詞で，部分に分けられない scheme, plan, system などを対象にする．

▶ Predictive schemes for reactivity of PAHs based on whole molecule properties such as the HOMO energy or ionization potential can fail badly for large PAHs.

▶ On the whole, the Rice–Ramsperger–Kassel–Marcus theory has proved very satisfactory.

これに対して，all は個々の構成要素を念頭におく．

whose

人と物に共通して用いられ，of which と同義である．

▶ Lanthanum combines energetically with hydrogen at 240°, forming a black product approximating LaH_3, but whose exact composition depends on temperature and the pressure of the hydrogen.

with

測定機器や方法を記する際，**…を用いて**の意味で使用する．

▶ UV–vis spectra were measured with Cary models 14 and 15 instruments.

measured の代わりに，determined, examined, obtained, taken などが，with の代わりに，by, using, by using も用いられる．

▶ Line positions were measured accurately by using field markers generated by an NMR gaussmeter while the microwave frequency was measured by a microwave frequency counter.

X

X-ray diffractometry

X線回折法による**結晶構造解析**には，次の事項が記載されていることが望ましい．

(1) 化学式および分子量 M_r または式量 fw（ローマン体で小文字の fw の後には mp の場合同様，等号＝を用いず，一字分あける）．

(2) 結晶系および空間群，必要ならば，標準の空間群の番号を付記する．なお，-1 は $\bar{1}$，$-h$ は $\bar{h}$ と記する．

(3) **格子定数**（単位は Å と°），いずれも標準偏差を付記する．標準偏差はかっこ（ ）に入れて記載し，大きさを表す数値との間は詰める．数値と Å との間はあけるが，数値と°との間は詰める．

(4) 単位格子の体積 V

(5) 測定温度

(6) 単位格子中の分子数 Z と密度の計算値 D_{calcd} ($g\,cm^{-3}$)

(7) 使用 X 線または波長（Å）および線吸収係数 μ (cm^{-1})

(8) 回折データ収集に用いた結晶の色，形，サイズなど

(9) 測定反射数，観測できた反射数，観測値として採用する基準など

(10) R_1, R_w, wR_2 などの値，すなわち

$$R_1 = \left[\frac{\sum \|F_o| - |F_c\|}{\sum |F_o|}\right] \text{ for } I > 2\sigma(I)$$

$$R_w = \left[\frac{\sum w(|F_o| - |F_c|)^2}{\sum wF_o^2}\right]^{1/2} \text{ for } F_o > 3\sigma(F_o)$$

$$wR_2 = \left\{\frac{w(F_o^2 - F_c^2)^2}{\sum w(F_o)^2}\right\}^{1/2} \text{ (all data)}$$

例えば

$C_{25}H_{35}CoNP_3$: fw 501.4, monoclinic, $P2_1/c$, $a = 9.385(3)$ Å, $b = 12.502(2)$ Å, $c = 23.190(4)$ Å, $\beta = 92.66(2)°$, $V = 2718.0(11)$ Å^3, $T = 293(2)$ K, $Z = 4$, $D_{calcd} = 1.225$ g/cm^{-3}, μ (Mo Kα) $= 0.820$ mm^{-1}, 8023 reflections measured, 6232 unique ($R_{int} = 0.020$). R_1 [I ≧ 2 σ(I)] $= 0.0515$, and wR_2 (all data) $= 0.1171$.

特定の位置の原子を指定するには，元素記号に番号を付記する．
　C2－C3 distance,　C2－C3－C4 angle

付表1 原子，分子に用いる物理量の記号とSI単位

用語	日本語名	記号	SI単位
Atomic mass	原子量	m_a	kg
Atomic mass constant	原子質量定数	m_u	kg
Atomic number	原子番号	Z	dimensionless
Bohr magneton	ボーア磁子	μ_B	$A\,m^2$, $J\,T^{-1}$
Bohr radius	ボーア半径	a_0	m
Decay constant	減衰定数，崩壊定数	λ	s^{-1}
Dissociation energy	解離エネルギー	E_d, D	J
Electric dipole moment	電気双極子モーメント	p, μ	C m
Electric polarizability of a molecule	分極率（分子）	α	$F\,m^2$, $J\,C^2\,m^2$
Eletron affinity	電子親和力	E_{ea}	J
Electron rest mass	電子の静止質量	m_e	kg
Electronic term	電子項	T_e	m^{-1}
Elementary charge (of a proton)	電気素量（陽子）	e	C, A s
g-factor	g因子	g	dimensionless
Ionization energy	イオン化エネルギー	E_i, I	J
Magnetic moment (of a particle)	粒子の磁気モーメント	μ, m	$A\,m^2$, $J\,T^{-1}$
Magnetogyric ratio	磁気回転比	γ	$s^{-1}T^{-1}$, $C\,kg^{-1}$
Mass number	質量数	A	dimensionless
Nucleon number	核子数	A	dimensionless
Neutron number	中性子数	N	dimensionless
Planck constant	プランク定数	h	J s
Planck constant/2π	プランク定数/2π	$\hbar$	J s
Proton number	陽子数	Z	dimensionless
Proton rest mass	陽子の静止質量	m_p	kg
Quadrupole moment	四極子モーメント	Q, Θ, eQ	$C\,m^2$
Quantum numbers	量子数		
Principal	主量子数	n	dimensionless
Electron orbital	電子軌道	l, L	dimensionless
Electron orbital component	電子軌道成分	m_l, M_L	dimensionless
Electron spin	電子スピン	s, S	dimensionless
Electron spin component	電子スピン成分	m_s, M_S	dimensionless
Total angular momentum	全角運動量	j, J	dimensionless
Total angular momentum component	全角運動量成分	m_j, m_J	dimensionless
Vibrational	振動	v	dimensionless
Nuclear spin	核スピン	I	dimensionless
Nuclear spin component	核スピン成分	M_I	dimensionless
Rotational constants	回転定数	A, B, C	m^{-1}
Rotational term	回転項	F	m^{-1}
Total term	項（全体）	T	m^{-1}
Unified atomic mass unit	統一原子質量単位	m_u	kg
Vibrational term	振動項	G	m^{-1}

付表2　化学反応速度論に用いる物理量の記号とSI単位

用語	日本語名	記号	SI単位
Activation energy	活性化エネルギー	E, E_a	J mol^{-1}
Boltzmann constant	ボルツマン定数	k, k_B	J K^{-1}
Collision cross section	衝突断面積	σ	m^2
Collision frequency factor	衝突頻度因子	z_{AB}	$\text{m}^3 \text{mol}^{-1} \text{s}^{-1}$
Collision frequency of a particle	粒子の衝突頻度	Z_A	s^{-1}
Enthalpy of activation	活性化エンタルピー	$\Delta^\ddagger H, \Delta H^\ddagger$	J mol^{-1}
Entropy of activation	活性化エントロピー	$\Delta^\ddagger S, \Delta S^\ddagger$	$\text{J mol}^{-1} \text{K}^{-1}$
Extent of reaction	反応進行度	ξ	mol
Gibbs energy of activation	活性化ギブスエネルギー	$\Delta^\ddagger G, \Delta G^\ddagger$	J mol^{-1}
Half-life	半減期	$t_{1/2}$	s
Internal energy of activation	活性化内部エネルギー	$\Delta^\ddagger U, \Delta U^\ddagger$	J mol^{-1}
Overall order of reaction	総括反応次数	n	dimensionless
Photochemical yield	光化学収量	ϕ	dimensionless
Pre-exponential factor	前指数因子	A	$(\text{m}^3 \text{mol}^{-1})^{n-1} \text{s}^{-1}$
Quantum yield	量子収量	ϕ	dimensionless
Rate constant (nth order)	速度定数（n次）	k	$(\text{m}^3 \text{mol}^{-1})^{n-1} \text{s}^{-1}$
Rate of concentration change of substance B	物質Bの濃度変化速度	v_B, r_B	$\text{mol m}^{-3} \text{s}^{-1}$
Rate of conversion	変換速度	$\dot{\xi}$	mol s^{-1}
Rate of reaction	反応速度	v	$\text{mol m}^{-3} \text{s}^{-1}$
Reaction cross section	反応断面積	σ_r	m^2
Relaxation time	緩和時間	τ	s
Scattering angle	散乱角	θ	rad
Stoichiometric coefficient	化学量論係数	ν	dimensionless
Temperature, absolute	絶対温度	T	K
Thermal energy	熱エネルギー	kT	J
Volume of activation	活性化体積	$\Delta^\ddagger V, \Delta V^\ddagger$	$\text{m}^3 \text{mol}^{-1}$

付表 3　電磁気学に用いる物理量の記号と SI 単位

用　語	日本語名	記号	SI 単位
Capacitance	キャパシタンス	C	F, $C\,V^{-1}$
Charge density	電荷密度	ρ	$C\,m^{-3}$
Conductance	コンダクタンス	G	S
Conductivity	伝導率	κ	$S\,m^{-1}$
Dielectric polarization	誘電分極	P	$C\,m^{-2}$
Electric charge	電荷（電気量）	Q	C, $A\,s$
Electric current	電流	I	A
Electric current density	電流密度	j, J	$A\,m^{-2}$
Electric dipole moment	電気双極子モーメント	p, p_e, μ	$C\,m$
Electric displacement	電気変位	D	$C\,m^{-2}$
Electric field strength	電場の強さ	E	$V\,m^{-1}$
Electric potential	電位	ϕ, V	V, $J\,C^{-1}$
Electric potential difference	電位差	U, $\Delta\phi$, ΔV	V
Electric resistance	電気抵抗	R	Ω
Electric susceptibility	電気感受率	χ_e	dimensionless
Electromotive force	起電力	E	V
Impedance	インピーダンス	Z	Ω
Inductance	インダクタンス	H	H, $A\,m^{-1}$
Magnetic field strength	磁場の強さ	H	H, $A\,m^{-1}$
Magnetic flux	磁束	$\boldsymbol{\Phi}$	Wb, $V\,s$, $J\,A^{-1}$
Magnetic flux density	磁束密度	B	T, $V\,s\,m^{-2}$
Magnetic induction	磁気誘導	B	T, $V\,s\,m^{-2}$
Magnetic moment	磁気モーメント	μ, m	$A\,m^2$, $J\,T^{-1}$
Magnetic susceptibility	磁化率	χ	dimensionless
Magnetization	磁化	M	$A\,m^{-1}$
Molar magnetic susceptibility	モル磁化率	χ_m	$m^3\,mol^{-1}$
Peameability	透磁率	μ	$H\,m^{-1}$
Permeability of vacuum	真空の透磁率	μ_0	$H\,m^{-1}$
Permittivity	誘電率	ε	$F\,m^{-1}$
Permittivity of vacuum	真空の誘電率	ε_0	$F\,m^{-1}$
Polarizability (of a particle)	分極率（粒子の）	α	$F\,m^2$, $C\,m^2\,V^{-1}$
Relative permeability	比透磁率	μ_r	dimensionless
Relative permittivity	比誘電率	ε_r	dimensionless
Resistance	抵抗	R	Ω
Resistivity	比抵抗	ρ	$\Omega\,m$
Self-inductance	自己インダクタンス	L	H
Surface charge density	表面電荷密度	σ	$C\,m^{-2}$
Voltage	電圧	U, $\Delta\phi$, ΔV	V, $J\,C^{-1}$
Work function	仕事関数	Φ	J

付表 4　電気化学に用いる物理量の記号と SI 単位

英語名	日本語名	記号	SI 単位
Charge number of a cell reaction	電池反応の電荷数	n, z	dimensionless
Charge number of an ion	イオンの電荷数	z	dimensionless
Conductivity	伝導率	κ	$S\,m^{-1}$
Diffusion rate constant	拡散速度定数	k_d	$m\,s^{-1}$
Electric current	電流	I	A
Electric current density	電流密度	j	$A\,m^{-2}$
Electric mobility	イオンの移動度	u	$m^2\,V^{-1}\,s^{-1}$
Electrochemical potential	電気化学ポテンシャル	$\tilde{\mu}$	$J\,mol$
Electrode potential	電極電位	E	V
Electrokinetic potential	界面動電位	ζ	V
Electrolytic conductivity	電解伝導度	κ	$S\,m^{-1}$
Electromotive force (emf)	起電力	E	V
Elementary charge	電気素量	e	C
Faraday constant	ファラデー定数	F	$C\,mol^{-1}$
Half-wave potential	半波電位	$E_{1/2}$	V
Ionic strength,	イオン強度	$I_m, (I_c)$	$mol\,m^{-3}$
Mass-transfer coefficient	物質移動係数	k_d	$m\,s^{-1}$
Mean ionic activity	平均イオン活量	$a_\pm$	dimensionless
Mean ionic molality	平均イオン質量モル濃度	$m_\pm$	$mol\,kg^{-1}$
Molar conductivity (of an electrolyte)	モル伝導率 （電解質の）	Λ	$S\,m^2\,mol^{-1}$
Molar conductivity (of an ion)	モル伝導度（イオンの）	λ	$S\,m^2\,mol^{-1}$
Overpotential	過電圧	η	V
pH	pH	pH	dimensionless
Standard electrode potential	標準電極電位	$E^\ominus$	V
Standard electromotive force (emf)	標準起電力	$E^\ominus$	V
Surface charge density	表面電荷密度	σ	$C\,m^{-2}$
Transport number	輸率	t	dimensionless

付表5　一般化学に用いる物理量の記号とSI単位

用語	日本語名	記号	SI単位
Amount concentration	濃度	c	mol m^{-3}
Amount of substance	物質量	n	mol
Atomic weight	原子量	A_r	dimensionless
Avogadro constant	アボガドロ定数	L, N_A	mol^{-1}
Concentration	濃度	c	mol m^{-3}
Concentration of substance B	物質Bの濃度	$c_B, [B]$	mol m^{-3}
Degree of dissociation	解離度	α	dimensionless
Extent of reaction	反応進行度	ξ	mol
Mass concentration	質量濃度	ρ, γ	kg m^{-3}
Mass fraction	質量分率	w	dimensionless
Molality	質量モル濃度	m	mol kg^{-1}
Molar mass	モル質量	M	kg mol^{-1}
Molar volume	モル体積	V_m	$\text{m}^3 \text{mol}^{-1}$
Molarity	モル濃度	M	mol L^{-1}
Mole fraction	モル分率	x, y	dimensionless
Molecular weight	分子量	M_r	dimensionless
Number concentration	粒子数濃度	C, n	m^{-3}
Number of entities	要素粒子の数	N	dimensionless
Partial pressure of substance B	物質Bの分圧	p_B	$\text{Pa}, \text{N m}^{-2}$
Pressure	圧力	p, P	$\text{Pa}, \text{N m}^{-2}$
Relative atomic mass	相対原子質量	A_r	dimensionless
Relative molecular mass	相対分子質量	M_r	dimensionless
Solubility of substance B	物質Bの溶解度	s_B	mol m^{-3}
Stoichiometric coefficient	化学量論係数	ν	dimensionless
Surface concentration	表面濃度	Γ	mol m^{-2}
Volume fraction	体積分率	ϕ	dimensionless

付表 6 力学に用いる物理量の記号と SI 単位

用　語	日本語名	記号	SI 単位
Angular momentum	角運動量	L	$\text{kg m}^2\text{ s}^{-1}\text{ rad}$, J s
Density	密度	ρ	kg m^{-3}
Energy	エネルギー	E	J
Force	力	F	N, kg m s^{-2}
Gravitational constant	重力定数	G	$\text{N m}^2\text{ kg}^{-2}$
Hamilton function	ハミルトン関数	H	J
Kinetic energy	運動エネルギー	E_k, T, K	J
Lagrange function	ラグランジュ関数	L	J
Mass	質量	m	kg
Moment of force	力のモーメント	M	N m
Moment of inertia	慣性モーメント	I	kg m^2
Momentum	運動量	p	kg m s^{-1}
Potential energy	ポテンシャルエネルギー	E_p, V, Φ	J
Power	仕事率	P	W
Pressure	圧力	p, P	Pa, N m^{-2}
Reduced mass	換算質量	μ	kg
Relative density	相対密度	d	dimensionless
Specific volume	比体積	v	$\text{m}^3\text{ kg}^{-1}$
Surface tension	表面張力	γ, σ	N m^{-1}, J m^{-2}
Torque	トルク	T	N m
Weight	重量	G, W	N
Work	仕事	w, W	J

付表 7　核磁気共鳴分光に用いる物理量の記号と SI 単位

用語	日本語名	記号	SI 単位
Bohr magneton	ボーア磁子	μ_B, β	$J\,T^{-1}$
Bohr radius	ボーア半径	a_0	m
Chemical shift δ	化学シフト	δ	dimensionless
Delay time	遅延時間	τ	s
Hyperfine coupling constant	超微細結合定数	a, A, T	Hz
Larmor angular frequency	ラーモア角振動数	ω_L	s^{-1}
Larmor frequency	ラーモア振動数	ν_L	Hz
Magnetogyric ratio	磁気回転比	γ	$s^{-1}\,T^{-1}$
Nuclear magneton	核磁子	μ_N	$J\,T^{-1}$
Quadrupole moment	四極子モーメント	Q, Θ	$C\,m^2$
Quantum numbers	量子数		
Principal	主量子数	n	dimensionless
Electron orbital	電子軌道	l, L	dimensionless
Electron orbital component	電子軌道成分	m_l, M_L	dimensionless
Electron spin	電子スピン	s, S	dimensionless
Electron spin component	電子スピン成分	m_s, M_S	dimensionless
Nuclear spin	核スピン	I	dimensionless
Nuclear spin component	核スピン成分	M_I	dimensionless
Relaxation time	緩和時間		
longitudinal	縦方向	T_1	s
transverse	横方向	T_2	s
Shielding constant	しゃへい定数	σ	dimensionless
Spin–spin coupling constant	スピン-スピン結合定数	J	Hz

付表 8 高分子化学に用いる物理量の記号と SI 単位

用語	日本語名	記号	SI 単位
Bulk modulus	体積弾性率	K	Pa, $N\,m^{-2}$
Complex permittivity	複素誘電率	ε^*	$F\,m^{-1}$
Electrophoretic mobility	電気泳動移動度	μ	$m^2\,V^{-1}\,s^{-1}$
Fracture strain	破断ひずみ	$\gamma_f,\ \varepsilon_f$	dimensionless
Fracture stress	破断応力	σ_f	Pa, $N\,m^{-2}$
Glass-transition temperature	ガラス転移温度	T_g	K
Modulus of elasticity	弾性率	E	Pa, $N\,m^{-2}$
Number-average molecular weight	数平均分子量	M_n	dimensionless
Tensile strength	引張り強さ	σ	Pa, $N\,m^{-2}$
Viscosity	粘性率	$\eta,\ \mu$	Pa s
Volume fraction	体積分率	V_f	dimensionless
Weight-average molecular weight	重量平均分子量	M_w	dimensionless
Yield stress	降伏応力	σ_y	Pa, $N\,m^{-2}$
Yield value	降伏値	σ_y	Pa
Youg's modulus	ヤング率	E	Pa

付表 9 放射に用いる物理量の記号と SI 単位

用語	日本語名	記号	SI 単位
Absorbance	吸光度	A	dimensionless
Absorptance	吸光率	α	dimensionless
Absorption coefficient	吸光係数,吸収係数		
linear decadic	線吸光係数（常用対数）	a	m^{-1}
molar decadic	モル吸光係数（常用対数）	ε	$m^2\,mol^{-1}$
Angle of optical rotation	旋光角	α	dimensionless, rad
Angular frequency	角振動数	ω	s^{-1}, $rad\,s^{-1}$
Circular frequency	角振動数	ω	s^{-1}, $rad\,s^{-1}$
Emissivity	放射率	ε	dimensionless
Emittance	放射率	ε	dimensionless
Frequency	振動数,周波数	ν	Hz, s^{-1}
Molar refraction	モル屈折	R_m	$m^3\,mol^{-1}$
Planck constant	プランク定数	h	J s
Planck constant/2π	プランク定数/2π	$\hbar$	J s
Radiant energy	放射エネルギー	Q, W, Q_e	J
Radiant intensity	放射強度	I, I_e	$W\,sr^{-1}$
Reflectance	反射率	ρ, R	dimensionless
Refractive index	屈折率	n	dimensionless
Speed of light	光速		
in vaccum	真空中	c_0	$m\,s^{-1}$
in a medium	媒質中	c	$m\,s^{-1}$
Stefan–Boltzmann constant	ステファン–ボルツマン定数	σ	$W\,m^{-2}\,K^{-4}$
Transmittance	透過率	τ	dimensionless
Wavelength	波長	λ	m
Wavenumber (in vaccum)	波数	$\tilde{\nu}$	m^{-1}

付表 10 空間と時間に用いる物理量の記号と SI 単位

用　語	日本語名	記号	SI 単位
Acceleration	加速度	a	$m\ s^{-2}$
Angular frequency	角振動数	ω	s^{-1}, $rad\ s^{-1}$
Area	面積	A, A_s, S	m^2
Cartesian space coordinates	直交空間座標	x, y, z	m
Characteristic time interval	特性時間間隔	τ, T	s
Relaxation time	緩和時間		
Time constant	時定数		
Circular frequency	角振動数	ω	s^{-1}, $rad\ s^{-1}$
Diameter	直径	d	m
Distance	距離	d	m
Frequency	振動数, 周波数	ν, f	Hz, s^{-1}
Height	高さ	h	m
Length	長さ	l	m
Length of arc	弧長	s	m
Path of length	行程	s	m
Plane angle	平面角	$\alpha, \beta, \gamma, \theta, \phi$	dimensionless, rad
Polar coordinates,	極座標	r, θ, ϕ	m, dimensionless
spherical polar coordinates			
Position vector	位置ベクトル	$\boldsymbol{r}$	m
Radius	半径	r	m
Solid angle	立体角	ω, Ω	dimensionless, sr
Speed	速さ	v, u, w, c	$m\ s^{-1}$
Thickness	厚さ	d, δ	m
Time	時間	t	s
Velocity	速度	$\boldsymbol{v, u, w, c, \dot{r}}$	$m\ s^{-1}$
Volume	体積	V	m^3

付表 11 熱力学に用いる物理量の記号と SI 単位

用語	日本語名	記号	SI 単位
Absolute activity	絶対活量	λ	dimensionless
Activity coefficient	活量係数		
mixtures	混合物	f	dimensionless
solutes	溶質	γ	dimensionless
Affinity of a reaction	反応の親和力	A	$\mathrm{J\ mol^{-1}}$
Characteristic temperature	特性温度	Θ	K
Chemical potential	化学ポテンシャル	μ	$\mathrm{J\ mol^{-1}}$
Cubic expansion coefficient	体膨張係数	α	$\mathrm{K^{-1}}$
Energy	エネルギー	E	J
Enthalpy	エンタルピー	H	J
Entropy	エントロピー	S	$\mathrm{J\ K^{-1}}$
Equilibrium constant	平衡定数	K°, K	dimensionless
on a concentration basis	濃度に基づくもの	K_c	$(\mathrm{mol\ m^{-3}})^{\Sigma\nu}$
on a pressure basis	圧力に基づくもの	K_P	$\mathrm{Pa}^{\Sigma\nu}$
on a molality basis	質量モル濃度に基づくもの	K_m	$(\mathrm{mol\ kg^{-1}})^{\Sigma\nu}$
Fugacity	フガシティー	$f, \tilde{p}$	$\mathrm{Pa,\ N\ m^{-2}}$
Gas constant	気体定数	R	$\mathrm{J\ K^{-1}\ mol^{-1}}$
Gibbs energy	ギブズエネルギー	G	J
Heat	熱	q, Q	J
Heat capacity	熱容量	C	$\mathrm{J\ K^{-1}}$
molar	モル熱容量	C_m	$\mathrm{J\ K^{-1}\ mol^{-1}}$
at constant pressure	定圧熱容量	C_P	$\mathrm{J\ K^{-1}}$
at constant volume	定容熱容量	C_V	$\mathrm{J\ K^{-1}}$
Helmholtz energy	ヘルムホルツエネルギー	A	J
Internal energy	内部エネルギー	U	J
Isothermal compressibity	等温圧縮率	κ	$\mathrm{Pa^{-1}}$
Joule–Thomson coefficient	ジュール—トムソン定数	μ, μ_{JT}	$\mathrm{K\ Pa^{-1}}$
Osmotic coefficient	浸透係数	ϕ	dimensionless
Osmotic pressure	浸透圧	Π	$\mathrm{Pa,\ N\ m^{-2}}$
Pressure	圧力	p, P	Pa
Pressure coefficient	圧力係数	β	$\mathrm{Pa\ K^{-1}}$
Ratio C_P/C_V	C_P と C_V 比	γ, κ	dimensionless
Relative activity	相対活量	a	dimensionless
Standard reaction enthalpy	標準反応エンタルピー	$\Delta_r H^\circ$	$\mathrm{J\ mol^{-1}}$
Surface tension	表面張力	γ, σ	$\mathrm{J\ m^{-2},\ N\ m^{-1}}$
Temperature	温度		
Celsius	セルシウス温度	θ, t	°C
thermodynamic	熱力学温度	T	K
Work	仕事	w, W	J

付表 12　輸送現象に用いる物理量の記号と SI 単位

用語	日本語名	記号	SI 単位
Coefficient of heat transfer	熱伝達係数	h	$\mathrm{W\,m^{-2}\,K^{-1}}$
Diffusion coefficient	拡散係数	D	$\mathrm{m^{-2}\,s^{-1}}$
Flux of a quantity X	量 X の流束	$J_x,\ J$	varies
Heat flow rate	熱流量	ϕ	$\mathrm{W,\ J\,s^{-1}}$
Heat flux	熱流束	J_q	$\mathrm{W\,m^{-2}}$
Kinematic viscosity	動粘性率	ν	$\mathrm{m^2\,s^{-1}}$
Mass flow rate	質量流量	$q_m,\ \dot{m}$	$\mathrm{kg\,s^{-1}}$
Mass-transfer coefficient	物質移動係数	k_d	$\mathrm{m\,s^{-1}}$
Thermal conductivity	熱伝導率	$\kappa,\ \lambda,\ k$	$\mathrm{W\,m^{-1}\,K^{-1}}$
Thermal diffusion coefficient	熱拡散係数	D_T	$\mathrm{m^2\,s^{-1}}$
Thermal diffusivity	熱拡散率	a	$\mathrm{m^2\,s^{-1}}$
Viscosity	粘性率	$\eta,\ \mu$	$\mathrm{Pa\,s}$
Volume flow rate	体積流量	$q_V,\ \dot{V}$	$\mathrm{m^3\,s^{-1}}$

索　引

●事項編

英字

a priori　Latin term (1)；roman type (1)
　ハイフン不使用　　hyphen (6a)
ab initio　Latin term (1)；roman type (1)
　ハイフン不使用　　hyphen (6a)
ad hoc　Latin term (1)；roman type (1)
　ハイフン不使用　　hyphen (6a)

Bragg 反射　crystal plane and direction (3)
cis-/trans-
　化合物名立体異性記述　capital letter (1b)；
　　cis,trans (2)；italic type (4)；title (10)
　ハイフンの使用　　cis,trans (2)
　ハイフン不使用　　cis,trans (1)
　文頭―非化合物名への　capital letter (2)；
　　cis,trans (1)；roman type (4)；

erythro-/threo-
　化合物名立体異性記述　capital letter (1b)；
　　italic type (4)
　文頭―非化合物名への　capital letter (2)；
　　roman type (4)
etc.　etc.

Hermann–Mauguin の記号　crystallographic
　groups and space groups
hyper　prefix (1)
hypo　prefix (1)

i-　化合物名接頭辞 iso の略記　abbreviation
　(11)；italic type (5)
in situ　Latin term (1)；roman type (1)
　ハイフン不使用　　hyphen (6a)
in vitro　Latin term (1)；roman type (1)

　ハイフン不使用　　hyphen (6a)
in vivo　Latin term (1)；roman type (1)
　ハイフン不使用　　hyphen (6a)
iso　化合物名接頭辞　abbreviation (11)；italic
　type (5)

macro　prefix (1)
mega　prefix (1)
Miller 指数　crystal plane and direction

n-　化合物名の補助記号　abbreviation (11)
nano　prefix (1)；units (20)

o-, m-, p-　化合物名の位置記号　capital letter
　(1b)；italic type (4)

Schönflies の記号　crystallographic groups and
　space groups；symmetry elements and point
　group
sec-　化合物名接頭辞　abbreviation (11)；
　capital letter (1b)；italic type (4)
semi　prefix (1)
SI 単位　付表 2-12
status quo　Latin term (1)；roman type (1)
super　prefix (1)
syn-/anti-
　化合物名立体異性記述　capital letter (1b)；
　　italic type (4)
　文頭―非化合物名への　capital letter (2)；
　　roman type (4)

tert-　化合物名接頭辞　abbreviation (11)；
　capital letter (1b)；italic type (4)

ultra　prefix (1)

via　Latin term (1)；via

索引　　　　　　　　　　　　　146

vice versa　　Latin term (1)；roman type (1)

wise　接尾辞　　adverb (3)

X線回折法
　　結晶構造解析に　　X-ray diffractometry (1–10)

ア 行

アステリスク　　asterisk；excited electronic state

イオン電荷　　ionic charge
イタリック体
　　学名　　capital letter (13)
　　化合物名の補助記号　　abbreviation (11)；capital letter (1b)；italic type (4)
　　関数の記号　　function (2)；italic type (6)
　　強調のために使用　　italic type (12)
　　元素記号による位置記号　　italic type (4)
　　最初の定義での使用　　italic type (13)；quotation mark (3)
　　雑誌巻数　　italic type (15)；references and note (7)
　　雑誌名，書籍名　　italic type (15)；references and note (7,9)
　　軸の記述　　axis；italic type (7)
　　物理量の記号　　italic type (2)；physical quantities (1,2)；subscript (1)
　　ベクトルは太字の　　boldface type (3)；vector
　　ベクトル成分　　italic type (8)；
　　変数の記述　　variable (1–3)
一般化学
　　物理量の記号とSI単位　　付表5
色の組み合わせ　　hyphen (7)
引用符　　quotation mark (1–4)
引用文献と注記　　references and notes (1–14)
　　学位論文の引用　　references and notes (10)
　　学術講演の引用　　references and notes (13)
　　雑誌巻数　　italic type (15)；references and note (7)
　　雑誌名の略記　　em dash (2)；italic type (15)；references and notes (1,3–5,7)
　　使用文字種　　references and notes (7)
　　書籍の引用　　italic type (15)；references and notes (9)
　　政府刊行物の引用　　references and notes (11)
　　注記の記述　　references and notes (14)
　　特許明細書の引用　　references and notes (12)
　　複数の著者　　references and notes (2)

複数の論文を引用　　references and notes (8)
上付き文字　　superscript
　　記述例　　subscript (3)
　　酸化数　　oxidation number (1,2)
　　同位体　　element (5)
　　熱力学での使用　　a, an (6a)；superscript
　　比旋光度　　specific rotation
　　表題中では避ける　　title (5)
　　文献の引用　　citation in text (1–3)

英国式・米国式の違い
　　ハイフンの使用　　prefix (1)；un (1)
　　分節法　　syllabication (2)
演算子　　equation (7)；symbols

大文字　　capital letter (1–13)
　α Particle, γ Ray　　capital letter (1a)
　　化合物名の頭字　　capital letter (4,5)
　　元素記号　　capital letter (12)
　　固有名詞／固有名詞由来形容詞　　capital letter (6,7)
　　参照図表，引用文献の章節　　capital letter (8,9)
　　商標　　capital letter (10,11)
　　にしない場合　　capital letter (1b,11,13)；title (7,9,14)
　　文頭の元素記号の後　　capital letter (3)

カ 行

化学結合，結合軌道
　π軌道，σ結合　　bond, bonding orbital (2)；Greek letter (3)
　　会合　　bond, bonding orbital (1)
　　水素結合　　bond, bonding orbital (1)
　　単結合　　bond, bonding orbital (1,2)
　　二重・三重結合　　bond, bonding orbital (1)
化学反応速度論
　　物理量の記号とSI単位　　付表2
化学方程式（化学反応式）　　chemical equation (1–4,6)
核磁気共鳴分光　　NMR spectroscopy
　　物理量の記号とSI単位　　付表7
核反応　　chemical equation (5)
学名
　　生物の科名以上および普通名詞扱い　　capital letter (13)
　　属名，種名　　capital letter (13)；italic type (14)；title (17)

147　　　　　　　　　　　　　　　　　　　　索　引

化合物名
　化学式との混用は避ける　　chemical name (3)
　化学式による記述　　chemical name (2)
　角かっこ　　bracket (3-5)
　化合物番号　　boldface type (2); number, numeral (19a,19b)
　化合物名の位置記号　　capital letter (1b); italic type (4)
　化合物名の補助記号　　abbreviation (11); capital letter (1b); italic type (4)
　化合物名立体異性記述　　capital letter (1b); cis,trans (2); italic type (4); title (10)
　元素記号による位置記号　　italic type (4)
　接頭辞　non-　と　chemical name (5)
　接尾辞　-like　と　chemical name (6)
　二語名称のハイフン不使用　　chemical name (4); hyphen (6)
　表題中の——　　title (16)
　ローマ体で記述　　chemical name (1)
過去分詞
　形容詞としての　　adjective (3)
可算名詞　　noun (1,2)
頭字語
　読み方　　abbreviation (10b)
かっこ　　bracket (1-6); parentheses (1-7)
　化合物名中の角かっこ　　bracket (2-4)
　関係式における濃度角かっこ　　bracket (1)
　結晶軸の角かっこ　　bracket (6); crystal plane and direction (1)
　数式の　　equation (8,2n)
　製造社名　　parentheses (6)
　同位体修飾化合物の　　bracket (5)
　丸かっこ　　parentheses (1-7)
関係詞
　関係形容詞　　restrictive clause, nonrestrictive clause
　関係代名詞　　that; which (1-2); whose
　関係副詞　　where (1-2)
　制限的用法　　restrictive clause, nonrestrictive clause (1)
　前置詞を伴う　　which (2)
　非制限的用法　　restrictive clause, nonrestrictive clause (2)
観察機器
　接尾辞　　scope
観察の
　接尾辞　　scopic
観察法
　接尾辞　　scopy

冠詞
　冠詞なし　　most; noun (3,4); the (4)
　冠詞不可欠　　same
関数の記号
　イタリック体　　function (2); italic type (6)
　三角関数，指数，対数　　equation (8)
　ローマ体　　equation (8); function (2); roman type (7)
脚注
　表の　　abbreviation (9); table (2,3,5)
　略記　　month; table (2,3)
キャプション　　boldface type (1); caption (1,2)
　→題目
　図やグラフの説明　　caption (1,2); colon (3)
　図面　　caption (1)
　表の　　caption (2); table (1,2)
行列の成分　　boldface type (3); italic type (8)
　→マトリックス
ギリシャ文字　　Greek letter (1-5)
　化合物名の位置記号　　Greek letter (4)
　結合軌道　　Greek letter (3); bond (2)
　ハイフンの使用　　Greek letter (5)
　物理量の記述　　Greek letter (1); physical quantities (1)
　放射線　　Greek letter (2)
記録法
　接尾辞　　graphy; metry
記録機器
　接尾辞　　graph; meter
記録図
　接尾辞　　gram
空間と時間
　物理量の記号と SI 単位　　付表 10
句動詞　　hyphen (3)
組立て
　synthesis とはしない　　synthesis (3)
形態学　　morphology
形容詞　　adjective (1-4)
　過去分詞の　　adjective (3)
　現在分詞の　　adjective (2)
　接辞と同形の　　counter (2); hetero (2); homo (2); like (5)
　接続詞ではない　　due to
　動詞と同形の　　disperse
　特定の前置詞を伴う　　adjective (4)
形容詞（複合）
　位置　　adjective (1)

索　引　　　　　　　　　　148

ギリシャ文字を冠した　　gamma ray；Greek letter (5)
形容詞への接辞　　like (5)；non (6)；prefix (5)
形容詞―形容詞の――　　hyphen (5f)
形容詞―名詞の――　　hyphen (5e)
三語以上の――　　hyphen (10)
数詞を含む　　hyphen (5h)
数と単位からなる――　　hyphen (17,18)；number, numeral (9)；units (15)
前置詞と抽象名詞の　　importance, of
中間語は複数形にしない　　hyphen (5h)
中間語を複数形とする例外　　hyphen (9)
ハイフン既存形容詞への接頭辞　　hyphen (11)；non (6)；prefix (5)
ハイフンの使用　　hyphen (5,7-13,15,17-20)
ハイフン不使用の例外　　hyphen (17)；newly, recently；units (15)
比較級, 最上級を含む――　　hyphen (9)
表題中の　　title (12)
副詞を含む形容詞　　hyphen (5g)；hyphen (8)
副詞句としての誤用　　based on；due to
複数の人名による　　en dash (3)
名詞―過去分詞の形容詞　　hyphen (5d)
名詞―形容詞の形容詞　　hyphen (5b)
名詞―現在分詞の形容詞　　hyphen (5c)
名詞―名詞の形容詞　　hyphen (5a)
ラテン語の――はハイフン不使用　　Latin term
結合軌道　　Greek letter (3)；bond (2)
結晶群, 空間群　　crystallographic groups and space groups；italic type (9)；symmetry elements and point group
結晶格子　　crystal lattice
結晶構造解析
　X線回折法のデータの記述　　X-ray diffractometry (1-10)
結晶面, 面指数, 結晶軸　　crystal plane and direction (1-3)
結論
　――の時制　　conclusion
現在分詞
　形容詞としての　　adjective (2)
　独立分詞　　comma (5)
原子軌道　　atomic orbital (1-3)
原子・分子　　atoms and molecules
　物理量の記号と SI 単位　　付表1
元素記号
　大文字の使用　　capital letter (3,12)；element (2)
　化合物名の位置記号　　italic type (4)
　化合物の位置記号以外　　capital letter (12)；

roman type (5)
不定冠詞　　a, an (6)；element (3)
名詞に冠する　　element (6)；hyphen (2)；roman type (5)
読み方　　a, an (6)；element (3)
元素名　　element (1)

口語形の不使用　　about, around (1) are, is (1-2)；cannot；generally
格子定数　　X-ray diffractometry (3)
合成　　synthesis (1)
高分子化学
　物理量の記号と SI 単位　　付表8
ゴシック体　　boldface type (1-4)
　引用文献の発行年　　boldface type (4)；references and notes (7)
　化合物番号　　boldface type (2)；number, numeral (19a,19b)
　表題の　　boldface type (1)
　ベクトル　　boldface type (3)；vector
固有名詞
　人名由来の単位名　　units (4)
　接頭辞がつく場合　　prefix (8)
　ハイフンの使用　　hyphen (4)；non (4-6)；prefix (8)
　ハイフン不使用　　hyphen (6b)；hyphen (12)
コロン　　colon (1-4)
　図の説明に　　colon (3)
　測定値の記載　　colon (2)
　比の記述　　colon (4)
　例証, 換言, 要約などの節に　　colon (1)
混合物, 固溶体　　en dash (4,5)；slash (1)
コンマ　　comma (1-16)
　引用文献の記述　　citation in text (1)
　引用文の前　　comma (14)
　語句の列記　　comma (1,4)
　五桁を超える数字　　comma (15)
　コンマの不使用　　comma (6)
　重文　　comma (11)
　挿入語句の前後　　comma (3,12)；however
　非制限的な句や節の前後　　comma (9)
　非制限的な従属接続詞の前　　comma (8)
　複文の従属節の後　　comma (7)
　分詞構文の切れ目　　comma (5)
　文頭の導入句の後　　comma (2)
　ラテン語の挿入　　e.g.；i.e.

サ　行

酸解離指数　p*K*a　　italic type (11)

酸化数
　元素記号につける　oxidation number (1,2)
　元素名につける　oxidation number (3)
参考文献　→引用文献と注記

紫外–可視分光
　測定値の記述　UV–visible spectroscopy
軸
　イタリック体，ハイフン不使用　axis
　回転軸，開映軸　symmetry elements and point group
　結晶軸　italic type (7)；bracket (6)
　座標軸　italic type (7)
指数，対数　equation (8)；function (1)
時制
　一般的な真理，習慣　past tense, present tense (5)
　経緯，経過　past tense, present tense (1)
　結果，考察，結論の　past tense, present tense (3)
　事実　past tense, present tense (2)
　──の一致　past tense, present tense (4)
下付き文字　subscript (1–3)
　pKa　italic type (11)
　記述例　subscript (3)
　元素の　element (4)
　熱力学の記号　subscript (2)
　比旋光度　specific rotation
　表題中では避ける　title (5)
　物理量の記号　iatlic type (2,3)；subscript (1)
質量数　mass number
集合名詞　a, an (3)；noun (2)
　couple　noun (2)
　group　noun (2)
　majority　noun (2)
　number　noun (2)
　pair　noun (2)
　series　noun (2)
　staff　noun (2)
　team　noun (2)
　variety　noun (2)
　冠詞　the (3)
　数と単位の組み合わせ　less；units (22)
　単数扱い　a, an (3)；majority (1)；series of, a；set of, a；units (22)；variety
　単数・複数の選択　noun (2)；singular and plural (4)
　複数扱い　a, an (3)；majority (1)；series of, a；set of, a；variety
終止符　period (1–4)

誌名の略記　reference and notes (1–8)
表題にはつけない　title (2)
表の題目に──は付けない　caption (2)
重文　comma (11)；or (4)
受動態，能動態　active voice, passive voice (1,2)
　受動態の主語に注意　fill；irradiate
　受動態は間違い　comprise
書籍
　→単行本
人名由来の下付き文字
　ローマン体で　italic type (3)
水和物
　分子式中で　centered dot (2)
数学的定数　roman type (6)
数学の記号
　定義は不要　symbols
数詞
　ハイフンの使用　hyphen (5h)；hyphen (15)
　ハイフン不使用　fold (1)
数字　figure；number, numeral (1–24)
　引用文献番号　number, numeral (18)
　大きな数　number, numeral (7)
　化合物番号　boldface type (2)；number, numeral (19a,19b)
　記述法　number, numeral (2–24)；roman type (8)
　時刻　number, numeral (17)
　数と単位の形容詞でハイフン使用　hyphen (17,18)；number, numeral (9)
　数と単位の形容詞でハイフン不使用　hyphen (17a–17d)；number, numeral (9a–9d)
　数の範囲　en dash (2)；number, numeral (22,24)
　正負号を伴う数の範囲　number, numeral (23)
　単位記号との間隔　number, numeral (2)
　綴り方　number, numeral (1)
　綴る数の範囲　number, numeral (10)
　年代の記述　number, numeral (15)
　日付　number, numeral (16)
　比と分数　number, numeral (12–14)
　文頭の数字　roman type (8)；number, numeral (1)
　本文中の節や文　number, numeral (20)
　本文中の図表　number, numeral (21)
数式　equation (1–12)
　値の単位の表示法　units (24)
　演算子はローマン体　roman type (6)
　記号の説明　where (2)

索　引

記述形式　　equation (7-12)
　項　　　term (1)
　別途改行して挿入する数式　　equation (3-6)
　本文中での使用　　equation (1,2)
スラッシュ　　slash (1-5)
　混合物の記述　　slash (1)
　接続詞 or の代用は不可　　slash (5)
　単位の記述　　slash (3,4)
　担持触媒　　en dash (5)
　比の記述　　slash (2)
　分数　　fraction (2,4)；slash (5)
製造
　synthesis とはしない　　synthesis (3)
製造社名
　表題には用いない　　title (5)
　本文中での記述　　parentheses (6)
製法
　synthesis とはしない　　synthesis (2)
赤外分光法
　測定値の記述　　IR spectroscopy
接続詞
　あるいは　　or (1)
　かどうか　　whether (1–2)
　コンマの不使用　　comma (6)
　従属接続詞　　comma (7)；that
　すると　　as (3)
　する時はいつでも　　whenever
　する所はどこでも　　wherever
　だから　　as (4)；because (1)
　であるが　　although；though
　ということ　　that
　等位接続詞　　and (1–6)；and/or；but；comma (6,11)
　ところが　　whereas
　と同様に　　as well as
　なのに　　while
　にもかかわらず　　though
　のように　　as (2)
　否定表現　　nor
　前にセミコロンを用いない　　semicolon (4)
　まで　　till, until
　よりも　　than
接頭辞　　prefix (1–9)
　10^n, 10^{-n} を意味する　　units (20)
　亜…　　sub (1,2)
　異種…　　hetero (1)
　下位…　　sub (1,2)
　化合物名への接頭辞　　prefix (6)
　擬…　　pseudo (1,2)

　逆…　　counter (1)
　共…　　co (1,2)
　形容詞と同形の　　counter (2)；hetero (2)；homo (2)
　後…　　post (1,2)
　固有名詞への接頭辞　　prefix (8)
　再…　　re (1,2)
　自動…　　auto (1,2)
　数字への接頭辞　　prefix (7)
　接頭辞後続語の省略の場合　　hyphen (19)；prefix (9)
　接頭辞の連続でハイフン使用　　non (5)
　前…　　pre (1,2)
　相互…　　inter (1,2)
　多…　　multi
　近い…　　peri (2)
　中間…　　mid (1,2)
　同種…　　homo (1)
　内部…　　inta (1,2)
　倍数接頭辞　　prefix (3,4)
　ハイフン既存の形容詞への　　hyphen (11)；non (6)；prefix (2b,2c,5)；pseudo (2)
　ハイフンの使用　　anti (2)；bi (2)；by-；chemical name (5)；co (2)；inter (2)；intra (2)；mid (2)；non (3,4)；peri (2)；pre (2)；prefix (2)；post (2)；pseudo (2)；re (2)；sub (2)；un (2)
　ハイフン不使用　　anti (1)；auto (1,2)；bi (1)；co (1)；counter (1)；hetero (1)；homo (1)；inter (1)；intra (1)；mid (1)；multi；non (1,2)；peri (1)；photo；post (1)；pre (1)；prefix (1)；pseudo (1)；re (1)；sub (1)；un (1)
　反…　　counter (1)；un (1,2)
　反対…　　anti (1,2)
　非…　　non (1–6)
　光…　　photo
　不…　　un (1,2)
　副…　　sub (1,2)
　副次的…　　by-
　二つ…　　bi (1,2)
　回りの…　　peri (1)
接尾辞　　suffix (1,2)
　…観察機器　　scope
　…観察の　　scopic
　…観察法　　scopy
　…記録法　　graphy；metry
　…記録機器　　graph；meter
　…記録図　　gram
　…形　　morph

…形態　　morphism
形容詞と同形の　　like (5)
…の形態をもつ　　morphic
…倍　　fold (1,2); hyphen (16); suffix (2)
ハイフンの使用　　fold (2); hyphen (16); like (1–5); suffix (b)
ハイフン不使用　　adverb (3); fold (1); suffix
…方向（wise）　　adverb (3)
…方法（wise）　　adverb (3)
…様の　　like (1–5)
…類似の　　chemical name (6); like (1–5); similar (2)
セミコロン　　semicolon (1–4)
引用文献の複数論文名　　references and notes (8)
測定値の列挙　　semicolon (2)
全角ダッシュ　　em dash (1)
特定の雑誌略名に対し　　em dash (2)
前置詞
　があるだけで　　except (2)
　がなければ　　except (2)
　からなる　　of (1)
　前置詞を伴う関係詞　　which (2)
　として　　as (1)
　とはちがって　　unlike
　に関しての　　about (2); of (2)
　によって　　by; with; on (2)
　の後で　　following (3)
　の上に　　above; on (1)
　のために　　because (2)
　の外は／を除いて　　except (1)
測定機器
　接尾辞　　graph; meter
測定記録図
　接尾辞　　gram
測定単位　　units (1–24)
　／で表す単位名　　units (17,18)
　／で表す単位記号　　units (13)
　10^n, 10^{-n} を意味する　　units (20)
　SI 基本単位　　units (7)
　SI 誘導単位　　units (7)
　記号の定義は不要　　units (5)
　人名由来の単位名は小文字で　　units (4)
　数式で示した値の単位の表示　　units (24)
　数字を伴う記述方法　　units (10–15)
　数字を伴わない――は綴る　　units (1)
　積で表す単位記号　　centerd dot; units (12)
　単位記号以外との組み合わせ　　units (17–19)

単位名と単位記号の混用は不可　　units (3)
単位名の複数形　　units (2)
単数扱い　　singular and plural (6)
動詞は単数形　　units (22)
範囲を示す数値がある場合　　units (21)
比較級は less を　　units (23)
表題, 題目中では綴る　　title (11); units (16)
略記しない, できない　　units (8,9)
ローマン体で　　roman type (3); units (6)
測定データ　　analytical data; data
NMR スペクトルの　　NMR spectroscopy
　→核磁気共鳴分光
X 線回折の　　X-ray diffractometry
コロンの使用　　colon (2)
紫外―可視スペクトルの　　UV–visible spectroscopy
質量スペクトルの　　mass spectrometry
赤外スペクトルの　　IR spectroscopy
測定値の列記　　semicolon (2)
定量分析の　　calculated; quantitative analysis
比旋光度の　　specific rotation
融点・沸点の　　melting and boiling points
測定法
　接尾辞　　graphy; metry

タ行

対称要素
　回転軸, 開映軸　　symmetry elements and point group
　対称心, 対称面, 点群　　symmetry elements and point group
対数　　equation (8); roman type (7)
代名詞　　this, that
　不定代名詞　　none
題目　　caption (1,2)
　図やグラフの説明　　colon (3)
　表の　　caption (2); table (1,2)
多成分系　　en dash (5); slash (1)
ダッシュ
　全角ダッシュ　　em dash (1)
　特定の雑誌略名に対し　　em dash (2)
　二分ダッシュ　　en dash (1–5)
単位記号　　→測定単位
単行本
　参照文献の　　boldface type (3); references and notes (9)
担持触媒　　en dash (5)
単数
　ics で終わる科学用語　　singular and plural (7)

一部省略した等位語列挙の最後の語　hyphen (19)
　個々を指示する語　singular and plural (8)
　集合名詞全体を問題　a,an (3); singular and plural (4)
　数詞-名詞の形容詞　hyphen (5h)
　測定単位　singular and plural (6); units (22)
　動詞は単数形　each (2); either (2); more than one; most (2); much of; neither of (2); number of (2); one of; set of, a; singular and plural (8); some (2); units (22)
　複数の名詞が一つの概念　and (2); each (1)
　単数・複数の選択/動詞の　singular and plural (1-11)
　前方語の数で決まる　as well as
　後方語の数で決まる　either (3); neither or (3); not only but also; or (1); singular and plural (3)
　強調したい内容で決まる　majority (1); none; series of, a; singular and plural (9)
　両方が可能　data; some (3)
　主語があるものの分数ないし一部　singular and plural (11)

抽象名詞
　冠詞　a, an (4); noun (4)

月の記述　month

定冠詞　the (1-5)
　つける場合　the (1-3)
　つけない場合　the (4)
　明確な話題への　a, an (1)
データ　data
　→測定データ
電気化学
　物理量の記号と SI 単位　付表 4
電子殻　atomic orbital
電磁気学
　物理量の記号と SI 単位　付表 3
電子状態　electronic state (1,2)
　原子の――　electronic state (1)
　分子の――　electronic state (2)
　励起電子状態　excited electronic state

同位元素, 同位体修飾
　かっこの使用　bracket (5); parentheses (4)
　読み方と不定冠詞　a, an (6b)
等号　equal (1,2)
　本文中での使用制限　equal (1)

ナ 行

中黒　centered dot (1-3)
　単位記号中の　centered dot (1)
　付加化合物, 水和物　centered dot (2)
　遊離基, イオンラジカル　centered dot (3); radical (1)

二重否定　double negative
二分ダッシュ
　紫外―可視分光　UV–visible spectroscopy
　数の範囲　en dash (2)
　対等な 2 語を結ぶ　en dash (1,3)

熱力学
　物理量の記号と SI 単位　付表 11

濃度　concentration (1-3)
　関係式中　concentration (2)
　複合形容詞で表す　concentration (3)
　質量モル濃度　concentration (1); italic type (10)
　モル濃度　italic type (10)
能動態, 受動態　active voice, passive voice (1-2)

ハ 行

倍数接頭辞
　ギリシャ語の　prefix (4a)
　ハイフンの使用　hyphen (5h); hyphen (16)
　ハイフン不使用　prefix (3,4)
　ラテン語の　prefix (4b)
ハイフンの使用　hyphen (1-20)
　色の組み合わせ　hyphen (7)
　化学物質名への接頭辞, 接尾辞　chemical name (5,6); like (3); prefix (6)
　ギリシャ文字を冠した名詞　Greek letter (5)
　形容詞―形容詞の形容詞　hyphen (5f)
　形容詞―名詞の形容詞　hyphen (5e)
　元素記号と名詞　element (6); hyphen (2)
　固有名詞の特例　hyphen (4)
　三語以上の形容詞　hyphen (10)
　数字と接尾辞　fold (2)
　数と単位からなる形容詞　hyphen (17,18); units (15)
　接頭辞　anti (2); bi (2); by–; chemical name (5); inter (2); intra (2); non (3-6); peri (2); post (2); pre (2); re (2)

索引

接頭辞後続語を省略の場合　　hyphen (19)
接尾辞に対して　　chemical name (6)；fold (2)；like (1-5)
等位語の列挙での一部省略　　hyphen (20)
ハイフン既存形容詞への接頭辞　　hyphen (11)；non (6)；prefix (5)
比較級，最上級を含む形容詞　　hyphen (9)
複合語　　hyphen (1)
複合動詞　　hyphen (3)
副詞を含む形容詞　　hyphen (5g)；hyphen (8)
分数　　fraction (1)
名詞―過去分詞の形容詞　　hyphen (5d)
名詞―形容詞の形容詞　　hyphen (5b)
名詞―現在分詞の形容詞　　hyphen (5c)
名詞―名詞の形容詞　　hyphen (5a)
例外的使用　　anti (2)；by-；co (2)；non (4)
ハイフン不使用
　cis,trans- 非化合物名　　cis,trans (1)
　句動詞　　hyphen (3)
　固有名詞と後の名詞との間　　hyphen (6b)；hyphen (12)
　軸の記述　　axis
　数詞と接尾辞　　fold (1)；prefix (3,4)
　数と単位の形容詞句での例外　　hyphen (17a-17d)；units (15)
　接頭辞に対して　　anti (1)；auto；bi (1)；co (1)；counter (1)；hetero (1)；homo (1)；inter (1)；intra (1)；non (1)；peri (1)；photo；post (1)；pre (1)；prefix (1)；re (1)
　接尾辞に対して　　fold (1)
　二語の化合物名　　chemical name (4)；hyphen (6)
　倍数接頭辞　　prefix (3,4)
　物質名の複合形容詞　　hyphen (6c)
　ラテン語の熟語　　Latin term；hyphen (6a)
範囲，数の　　en dash (2)；number, numeral (22,24)
番号付け
　説明番号　　parentheses (5)
　文献番号　　parentheses (7)

比較級，最上級
　形容詞の　　comparative and superlative adjevtives (1-5)；fewer；less；units (23)
　測定単位を伴う数値の　　units (23)
　副詞の　　comparative and superlative adverbs (1-3)；less
非制限的な従属節　　comma (8)；restrictive clause, nonrestrictive clause (2)；which

比旋光度　　specific rotation
日付・時間の記述　　number, numeral (16)
　略記しない　　month；units (8)；week
比の記述　　colon (4)；slash (2)；ratio
微分　　equation (9)
表　　table (1-5)
　記述法　　table (2,4)
　脚注　　table (3,5)
　題目　　caption (1)；table (1)
表題　　title (1-17)
　mg は Milligram と綴る　　title (11)
　意味の少ない語句は無用　　title (3,4)
　上付き，下付き文字は避ける　　title (5)
　会社名，商標は含めない　　title (5)
　学名の種名は常に小文字　　title (17)
　化合物名の補助記号　　title (10)
　キーワードを含める　　title (1)
　句動詞は副詞部分も頭字を大文字　　title (8)
　複合形容詞は先頭語の頭字のみ大文字　　title (12)
　接続詞，冠詞，前置詞は小文字　　title (7,7a,7b)
　先頭に The は不要　　title (2)
　単位記号は綴る　　units (16)
　ハイフン付きの接頭辞は小文字　　title (14)
　複合語は各成分の頭字を大文字　　title (13)
　名詞，代名詞，動詞，形容詞，副詞はキャピタライズ　　title (6)
　有機化合物体系名，ポリマー名　　title (15,16)
　略語・記号の回避　　abbreviation (2)；title (5)

付加化合物
　――の分子式　　centered dot (2)
不可算名詞　　graphy；metry；morphism；morphology；noun (3,4)；scopy
　attention　　and (2)
　equipment　　and (4)
　light　　complex, complicated
　spectroscopy　　both…and
　stability　　contrast (1)
　接尾辞による――　　morphism；scopy
　単数扱い　　some (2)
　定冠詞なし　　the (4)
　――との組み合わせ語　　part (2)
　不定冠詞なし　　a, an (4)
複合形容詞　　→形容詞（複合）
複合語
　形容詞―形容詞の――　　hyphen (5f)
　形容詞―名詞の――　　hyphen (5e)
　元素記号と名詞　　element (6)；hyphen (2)
　三語以上の形容詞　　hyphen (10)

索　引

154

ハイフンの使用　　hyphen (1,5)
比較級，最上級を含む形容詞　　hyphen (9)
複合動詞　　hyphen (3)
副詞を含む形容詞　　hyphen (5g)；hyphen (8)
名詞―過去分詞の形容詞　　hyphen (5d)
名詞―形容詞の形容詞　　hyphen (5b)
名詞―現在分詞の形容詞　　hyphen (5c)
名詞―名詞の形容詞　　hyphen (5a)
副詞　　adverb (1-3)
　形容詞からの派生　　adverb (1)
　形容詞とは異なる意味　　adverb (2)
　接尾辞 wise をつけたもの　　adverb (3)
複数
　集合名詞の成分を問題　　a, an (3)；singular and plural (4)
　多数を指示する語　　singular and plural (10)
　動詞は複数扱い　　number of (1)；series of, a；and (1)；singular and plural (2,10)；some (1)
　略語の　　abbreviation (12)；highest occupied molecular orbital
複文
　コンマの使用　　comma (7)
　制限的複文　　comma (10)
　非制限的複文　　comma (8,9)
普通名詞　　a, an (1)；noun (1)
　単数の不定冠詞　　a, an (1)
物質名詞　　a, an (4)；noun (3)
沸点　　melting and boiling points
物理定数
　イタリック体表記，ギリシャ文字表記　　constant；italic type (3)
物理量
　イタリック体表記，ギリシャ文字表記　　Greek letter (1)；physical quantities (1)
　記号と SI 単位　　付表 1-12；physical quantities (4)
　記号，変数はイタリック体　　italic type (2)；physical quantities (1,2)
不定冠詞　　a, an (1-6)
　頭字語の　　abbreviation (10b)
　元素記号の前　　a, an (6)
　集合名詞に　　a, an (3)
　普通名詞単数への　　a, an (1)
　母音字の前　　a, an (5)
　略語の　　a, an (5)；abbreviation (10a)
文献の引用
　本文中での　　citation in text (1-3)
分子軌道
　最高被占軌道　　highest occupied molecular orbital
分詞構文　　comma (5)
主文の能動態，受動態との一致　　active voice, passive voice (2)
分子量　　molecular weight
式量　　molecular weight
重量平均分子量　　molecular weight
数平均分子量　　molecular weight
相対分子量　　molecular weight
分数
　記述法　　fraction (2,4)；hyphen (15)；slash (5)
分節法　　syllabication (1-7)
　行末での送り箇所　　syllabication (3,4)
　固有名詞の分割は望ましくない　　syllabication (7)
　数字は分割できない　　syllabication (6)
　単音節語は分割できない　　syllabication (2)
　略語は分割できない　　syllabication (5)
ベクトル　　boldface type (4)；italic type (8)；vector
変数
　イタリック体　　italic type (2)；variable (1-3)
放射
　物理量の記号と SI 単位　　付表 9
放射線
　X 線　　alpha particle；capital letter (1a)；title (9)；X-ray diffractometry
　アルファ粒子　　alpha particle；capital letter (1a)；Greek letter (2)；title (9)
　ガンマ線　　gamma ray；capital letter (1a)；Greek letter (5)；title (9)
　紫外線　　UV-visible spectroscopy
　赤外線　　IR spectroscopy
　ベータ線　　beta particle；Greek letter (2)；title (9)
補足説明
　かっこの使用　　parentheses (1-3)
　非制限的例示　　such as (3)

マ　行

マトリックス　　boldface type (3)；italic type (8)
ミラー指数　　Miller indices；crystal plane and direction

索引

名詞
　可算名詞　　noun (1,2)
　集合名詞　　a, an (3)；noun (2)
　抽象名詞　　a, an (4)；noun (4)
　特定の前置詞をともなう　　noun (5)
　不可算名詞　　noun (3,4)
　普通名詞　　a, an (1)；noun (1)
　物質名詞　　a, an (4)；noun (3)
　――への不定冠詞　　a, an (4)；noun (4)

ヤ　行

矢印
　化学反応中の　　arrows in reactions

有機原子団
　定義不要の記号　　abbreviation (11)
融点　　melting and boiling points
遊離基
　陰イオンラジカル　　centered dot (3)；radical (2)
　化学式中で　　centered dot (3)；radical (1)
　陽イオンラジカル　　centered dot (3)；radical (2)
輸送現象
　物理量の記号とSI単位　　付表12

ラ　行

ラジカル　　→遊離基
ラテン語　　Latin term (1)
ラテン語略記
　ローマン体表記　　Latin term (2)；roman type (2)

力学
　物理量の記号とSI単位　　付表6
略語　　abbreviation
　一般に使用可能な　　別表
　引用文献中の　　references and notes
　測定単位の　　units (4,7)

定義　　abbreviation (4,7,8,9)
日付　　month
表題中の回避　　abbreviation (2)；title (5)
表中のみの略語　　table (3)
複数形　　abbreviation (12)
物理量の　　付表1–12
不定冠詞のつけ方　　a, an (5)；abbreviation (10)
有機原子団の　　abbreviation (11)
読み方　　a, an (5)；abbreviation (10)

励起状態　　asterisk；excited electronic state
例示　　e.g.

ローマン体
　化合物名　　chemical name (1)
　化合物名 iso　　abbreviation (11)
　関数の記号　　equation (8)；function (1)；roman type (7)
　元素記号　　capital letter (12)；element (6)；roman type (5)
　数学的定数　　roman type (6)
　数値　　roman type (8)
　生物の科名以上および普通名詞扱い　　capital letter (13)
　測定単位　　units (6)
　単位記号　　roman type (3)；physical properties (4)；units (6)
　熱力学の下付き記号　　subscript (2)
　分子量の記述　　molecular weight
　文頭非化合物名の Cis/Trans　　capital letter (2)
　文頭非化合物名の Erythro/Threo　　capital letter (2)
　文頭非化合物名の Syn/Anti　　capital letter (2)
　ラテン語　　Latin term (1,2)；roman type (1)
　ラテン語略記　　Latin term (2)；roman type (2)
　量子状態　　roman type (9)

●表現編

ア 行

間に（三者以上の間）	among
間に（二者の間）	between
扱う	deal with
当てる	apply (2)
あるいは	or (1)
以下のとおり	follow (2)
いずれか一方	either (3)
いずれも…ない	each (3)
一致する	correspond to；with
一般に	commonly；general, in；generally；usually
一方では	while；whilst
いろいろ	variety
影響を及ぼす	affect, effect
得る	get, obtain
多くの	much of
起こす	give rise to
同じ	same (1)
おのおの	each (1-3)；respectively
帯びる	assume
思われる	seem
およそ（数式中で）	symbols
およそ（文章中で）	about (1)；別表
終わる	result (2)

感光する	exposing
簡単な形にする	reduce (1)
関連させる	involve
起因する	due to；result (1)
帰すべきで	due to
基礎・根拠	basic, basis
基礎を置く	based on
帰着する	result (2)
基本的な	basic (3)
逆の	counter
共通に	commonly
興味がある	interest, in
比べて	while；whilst
結果として生じる	affect, effect
→果たす，遂げる	
決して…でない	means (3)
原因・根拠が…にある	lie in
合成	synthesis (1)
構成する	compose
構成要素の一つ	part
考慮する	account (2)
事柄	matter
異なった	differ from；different from；different to
このように	thus
コロイド粒子からなる	disperse

カ 行

があるだけで	except (2)
該当する	correspond to, with
仮定する	assume
かどうか	whether (1-2)
がなければ	except (2)
かなり	rather (1)；somewhat
可燃性の	flammable, inflammable
可能とする	allow (1)
からなる	compose；comprise；consist of；of (2)
関係する	involve
還元する	reduce (2)

サ 行

最近	newly；recently
最初の／最初に	first
させる	make
酸化させる，酸化する	oxidize
しかし	although；but；however
事実に反する記述	if…were
したがって	therefore (1-3)；thus
習慣的にいつも	usually
重要である	importance, of
手段，方法	means (1)
照会させる	refer to

照射する	irradiate
処理する	deal with
少しずつ	amount (2)
捨てる	comma (11)
すなわち	i.e ; namely ; or ; that is
する間に, するうちに	while
することにおいて	in…ing
すると	as (3) , on…ing
する時にはいつでも	whenever
する所の	where
する所はどこでも	wherever
正確に同じ	identical
製造	synthesis (2)
前者…, 後者…	former, the, latter, the
全体の	ole
相応する, 相当する	correspond to, with
そのうえ	as well
その他	etc.
それぞれ	respectively
それどころか	contrary
それほど	so (1)
それゆえ	hence ; so (3) ; therefore (1–3)

タ 行

対照する, 対比する	contrast
対, 対比して	versus, vs.
大多数・大部分の	majority ; part (4)
たいていの	most (1)
だから	as (4) ; because (1) ; since ; so (3)
だけでなく…も	as well as ; not only but (also)
だけれども	though
多少の	rather (1) ; somewhat
多数の	number of
例えば	i.e ; that is
多様の	variety
多量の	most (2)
貫いて	through
であるが	although ; though →しかし
であればあるほど	the (5)
提供する	provide (1)
適用する	apply (1)
できる	may, might (1)
手放す	comma (11)
照らす	irradiate
ということ	that
という条件で	provide, provided that (2)
同一の	same (1)
同様な, 同種の	similar (1)
同様に	as well as
通って	through ; via
とおなじ	same (2)
特殊の	particular
遂げる	affect, effect
ところが	whereas
として	as (1)
どちらも	either (1,2)
伴う	involve
執る	assume
どんな…でも	whatever

ナ 行

など	etc.
なのに	while ; whilst
なほど…である	so (2)
に関して	about (2) ; of (2) ; as to ; respect to, with ; term (2)
に関する限り	as for, as to
に従って	according to (1)
に次いで	follow (3)
についての	about (2) ; as for, as to (2) ; respect to, with
に続く	follow (2)
にほかならない	nothing other than
にもかかわらず	despite ; nevertheless ; nonetheless ; regardless of ; spite ; though
によって	by ; means (2) ; on (2) ; with
によれば	according to (2)
の後で	follow (3)
の上で	above ; on (1)
の結果として起こる	follow (1)
の見地から	term (2)
のことを指す	refer to
のために	because (2) ; owing to
の点から	term (2)
の外は／除いて	except (1)
のもとである	give rise to
のような	similar (2) ; such as (1–3) ; such... as
のように	as (2)

ハ 行

果たす　　affect, effect
反対の, 反対に　　contrary

比較して　　relative to
比較する　　compare, comparison (1-3)
比較的　　comparatively；relatively
光が…に当たる　　fall on
光が通る　　pass
光に当てる　　exposing
引き起こす　　bring about
　→もたらす
非常な／非常に　　extreme, extremely；so (1)
非常に…なので　　so (2)
一組の　　set of, a
一続きの　　series of, a
一つの　　one of
一つ（一人）もない　　no
一つより多い　　more than one
比率　　proportion
比例して　　as for, as to (2)；relative to

複雑な　　complex, complicated
含めて, 含む　　including
普通の, 普通に　　ordinary, ordinarily
部分的に　　partially；partly
分散する／させる　　disperse

経て　　through；via

包含する　　involve
放置する　　allow (3)
放冷する　　allow (3)
ほかの条件が同じなら　　other things being equal
ほとんど…ない　　hardly

マ 行

まで　　till, until

見込む　　allow (2)
水のような　　aqueous
　→水を含む, 水性の
満たす　　fill
見つける　　find, discover

むしろ　　rather (2)

明白な　　obvious
めいめいの　　each (1-3)；respectively

もう一つの　　another (1-2)
燃えやすい　　flammable, inflammable
もし…とすれば　　provide, provided that (2)
もしかしたら…であろう　　may, might (2)
用いて　　by；means (2)；with
基づいて　　based on；basis (2)
もまた　　also；as well as；too
　→同様に
問題　　matter

ヤ 行

やや　　rather (1)；somewhat

唯一の　　only

より大きい　　major
より重要な　　major
より少数の　　fewer
より少ない　　less
よりも　　than

ラ 行

らしい　　apparently；seem

理由を説明する　　account (1,3)；as (4)；because (1)；reason
　→それゆえ, だから
両者の, 両方の　　both…and
量の扱い　　amount (1)

連続した　　series of, a

論じる　　deal with

ワ 行

割合　　part (3)
割合に　　comparatively；relatively
割り当てられた部分　　portion

MEMO

MEMO

編著者略歴

松 永 義 夫（まつなが・よしお）

1929 年　岐阜県に生まれる
1952 年　東京大学理学部化学科卒業
1966 年　北海道大学理学部教授
現　在　北海道大学名誉教授
　　　　理学博士
主な著書　『物性化学』（裳華房）
　　　　　『現代の物理化学』（三共出版）
　　　　　『入門化学熱力学』（朝倉書店）

化学英語のスタイルガイド　　　定価はカバーに表示

2006 年 2 月 15 日　初版第 1 刷

監　修　社団法人 日本化学会
編著者　松　永　義　夫
発行者　朝　倉　邦　造
発行所　株式会社 朝　倉　書　店
　　　　東京都新宿区新小川町 6-29
　　　　郵便番号 １６２-８７０７
　　　　電　話　03（3260）0141
　　　　Ｆ Ａ Ｘ　03（3260）0180
　　　　http://www.asakura.co.jp

〈検印省略〉

© 2006〈無断複写・転載を禁ず〉　　新日本印刷・渡辺製本

ISBN 4-254-14073-8　C3043　　　　　Printed in Japan

核融合科学研 廣岡慶彦著	著者の体験に基づく豊富な実例を用いてプレゼン英語を初歩から解説する入門編。学会・会議に不可欠なコミュニケーションのコツを伝授。〔内容〕予備知識／準備と実践／質疑応答／国際会議出席に関連した英語／付録（予備練習／重要表現他）
理科系のための 入門英語プレゼンテーション 10184-8 C3040　　A 5 判 136頁 本体2500円	
核融合科学研 廣岡慶彦著	豊富な実例を駆使してプレゼン英語の実際を解説。質問に答えられないときの切り抜け方など、とっておきのコツも伝授する。〔内容〕心構え／発表のアウトライン／研究背景・動機の説明／研究方法の説明／結果と考察／質疑応答／重要表現
理科系のための 実戦英語プレゼンテーション 10182-1 C3040　　A 5 判 144頁 本体2700円	
核融合科学研 廣岡慶彦著	国際会議や海外で遭遇する諸状況を想定し、円滑な意思疎通に必須の技術・知識を伝授。〔内容〕国際会議・ワークショップ参加申込み／物品注文と納期確認／日常会話基礎：大学・研究所での一日／会食でのやりとり／訪問予約電話／重要表現他
理科系のための 状況・レベル別英語コミュニケーション 10189-9 C3040　　A 5 判 136頁 本体2700円	
核融合科学研 廣岡慶彦著	英文法の基礎に立ち返り、「英語嫌いな」学生・研究者が専門誌の投稿論文を執筆するまでになるよう手引き。〔内容〕テクニカルレポートの種類・目的・構成／ライティングの基礎的修練法／英語ジャーナル投稿論文の書き方／重要表現のまとめ
理科系のための 入門英語論文ライティング 10196-1 C3040　　A 5 判 128頁 本体2500円	
鹿児島大 中山　茂著	慣用的な表現法や語句、質疑応答も含めた実践的な要領やテクニック、英語でのメモのとり方、発表器材の活用法、原稿作成法など、例題により実践力を強化する。〔内容〕機能英語による口頭発表／機能英語による質疑応答／効果的な口頭発表
科学者のための 英語口頭発表のしかた 10082-5 C3040　　A 5 判 208頁 本体2900円	
M.アレイ著　静岡理工科大 志村史夫編訳	読者に情報を与え、納得させるという究極の目的を果す科学・技術文書とは。〔内容〕構成（整理・推移・詳述、強調）／語句（正確さ・明確さ・率直さ・親しみ・簡潔さ・流麗さ）／図表／通信文／取扱説明書／口頭発表／成功への手入れ／実行
理 科 系 の 英 文 技 術 10151-1 C3040　　A 5 判 248頁 本体3900円	
H.S.ロバーツ著　CSK 黒川利明・黒川容子訳	科学論文は、理論性、正確さ、事実を整理する能力が必要である。本書は初心者を対象に、犯しやすいまちがいを例示しながら、優れた作文技術の習得までを解説。〔内容〕小論文の書き方／報告書を書く／技術作文の道具／文体／就職の作文
科 学 英 文 作 成 の 基 本 10162-7 C3040　　A 5 判 164頁 本体3200円	
M.J.カッツ著　前神奈川大 桜井邦朋訳	科学論文を作り上げる過程を順に段階を追って具体的に記述。〔内容〕科学論文とは何か／論文を書く理由／自分にあった形式の選択／コンピュータで書く／言語／数値／材料と方法／なま データの組み立て／結果／図／脚注と付録／結論／他
科 学 英 語 論 文 の 基 礎 作 法 10073-6 C3040　　A 5 判 152頁 本体2600円	
前神奈川大 桜井邦朋著	論文を書く前に注意すべき事柄や具体的作業を指導。〔内容〕基礎編（常識の誤り、日本語の論理・英語の論理、論文を書く前に、論文を作る、例文でみる英語論文、私の経験から、他）／演習編（文章研究、まちがいやすい用法、他）／総括編
科 学 英 語 論 文 を 書 く 前 に 10068-X C3040　　A 5 判 200頁 本体3200円	
D.ビア・D.マクマレイ著 CSK 黒川利明・黒川容子訳	自己表現を高めるための具体的な方法を詳しく解説。〔内容〕エンジニアと作文／上手な技術作文のための指針／作文に散発するノイズをなくす／一般的技術文書／技術報告書／技術情報の入手／口頭発表／技術職につくために／コンピュータ利用
英 語 技 術 文 書 の 作 法 10150-3 C3040　　A 5 判 248頁 本体3400円	

岡山大 河本　修・アラバマ大 C.アレクサンダー，Jr.著	本書は科学論文の流れと同じ構成とし，単語や語句のみではなく主語と動詞からなる1500に及ぶ文全体を掲載。単語や語句などの表現要素を置き換えれば望む文章が作成可能で，より短時間で簡単に執筆できることを目指している。

実用的な英語科学論文の作成法

10193-7 C3040　　　　　A5判 260頁 本体3800円

黒木登志夫・F.H.フジタ著	科学者が日常出会うあらゆる場面を想定し，多くの文例を示しながら正しい英文手紙の書き方を解説。必要な文例は索引で検索。〔内容〕論文の投稿・引用／本の注文／学会出席／留学／訪問と招待／奨学金申請／挨拶状／証明書／お詫び／他

科学者のための 英文手紙の書き方（増訂版）

10038-8 C3040　　　　　A5判 224頁 本体2900円

井上信雄・E.E.ダゥブ著	日米の工学者が，自らの豊富な経験をふまえて著した，学生・技術者・研究者の必携書。〔内容〕一般的注意事項／執筆計画のたて方／論文にとりかかる前の準備／下書きの作り方／よい英文の書き方／最後の仕上げ／清書する場合の注意事項／他

英語技術論文の書き方

20035-8 C3050　　　　　A5判 180頁 本体2900円

J.ゾーベル著　CSK 黒川利明・黒川容子訳	計算機科学・数学的内容を含む論文やレポートの文体を解説し，発表にまで言及した入門書〔内容〕論文／文体：一般的ガイドライン／文体：具体的なこと／句読点／数学／グラフ，図，表／アルゴリズム／仮説と実験／編集／査読／短い講演

コンピュータサイエンスの 英語文書の書き方

10173-2 C3040　　　　　A5判 192頁 本体3200円

長岡技科大 若林　敦著	レポートや論文の作成に必要な作文技術の習得をめざし，豊富な文例と練習問題を盛り込んだ実践的なテキスト。独習用としても最適。〔内容〕事実と意見(区別する，書きわける)／わかりやすく簡潔な表現(文の三原則，文と文とのつなぎ方)

理工系の日本語作文トレーニング

10168-6 C3040　　　　　A5判 180頁 本体2800円

九工大 栗山次郎編著	"理系学生の実状と関心に沿った"コンパクトで実用的な案内書。〔内容〕コミュニケーションと表現／ピタゴラスの定理の表現力／コンポジション／実験報告書／レポートのデザイン・添削／口頭発表／インターネットの活用

理科系の日本語表現技法

10160-0 C3040　　　　　A5判 184頁 本体2600円

M.F.モリアティ著　前自治医大 長野　敬訳	「書く」ことは「考える」ことだ。学生レポートからヒポクラテスまで様々な例文を駆使し，素材を作品に仕上げていく方法をコーチ。〔内容〕科学を考える・書く／読者と目的／抄録／見出し／論文／図表／展望／定義／文脈としての分類／比較／他

「考える」科学文章の書き方

10172-4 C3040　　　　　A5判 224頁 本体3600円

冨田軍二著　小泉貞明・石舘　基補訂	自然科学関係の研究・調査・観察・観測等に従事する学生・研究者の必携書。〔内容〕序論／論文にまとめるまで／論文の構成／文章論／文献／図表／特殊事項の表示形式／原稿の仕上げと校正／付録：欧文雑誌略名一覧・英語論文文範集／他

新版 科学論文のまとめ方と書き方

10011-6 C3040　　　　　A5判 224頁 本体3800円

元室蘭工大 傳　遠津著 化学者のための基礎講座1	広くサイエンスに学ぶ人が必要とする英文手紙・論文の書き方エッセンスを例文と共に解説した入門書。〔内容〕英文手紙の形式／書き方の基本(礼状・お見舞い・注文等)／各種手紙の実際／論文・レポートの書き方／上手な発表の仕方等

科学英文のスタイルガイド

14583-7 C3343　　　　　A5判 192頁 本体3600円

元岡山理大 砂原善文著	研究発表・講演の準備からその実際の場面まで著者の永年の経験を生かして失敗例も混じえながら説得力のある魅力あふれた方法を伝授する。〔内容〕講演資料の作成(スライド，OHP)／講演原稿／朝食会／ジョークの種／フライトスケジュール

科学者のための 研究発表のしかた

10046-9 C3040　　　　　A5判 128頁 本体2400円

D.M.コンシディーヌ編
今井淑夫・中井 武・小川浩平・
小尾欣一・柿沼勝己・脇原将孝監訳

化 学 大 百 科

14045-2 C3543　　B5判 1072頁 本体58000円

化学およびその関連分野から基本的かつ重要な化学用語約1300を選び、アメリカ、イギリス、カナダなどの著名化学者により、化学物質の構造、物性、合成法や、歴史、用途など、解りやすく、詳細に解説した五十音配列の事典。Encyclopedia of Chemistry(第4版, Van Nostrand社)の翻訳。〔収録分野〕有機化学／無機化学／物理化学／分析化学／電気化学／触媒化学／材料化学／高分子化学／化学工学／医薬品化学／環境化学／鉱物学／バイオテクノロジー／他

くらしき作陽大 馬淵久夫編

元 素 の 事 典

14044-4 C3543　　A5判 324頁 本体7800円

水素からアクチノイドまでの各元素を原子番号順に配列し、その各々につき起源・存在・性質・利用を平易に詳述。特に利用では身近な知識から最新の知識までを網羅。「一家庭に一冊、一図書館に三冊」の常備事典。〔特色〕元素名は日・英・独・仏に、今後の学術交流の動向を考慮してロシア語・中国語を加えた。すべての元素に、最新の同位体表と元素の数値的属性をまとめたデータ・ノートを付す。多くの元素にトピックス・コラムを設け、社会的・文化的・学問的な話題を供する

前学習院大 髙本 進・前東大 稲本直樹・
前立教大 中原勝儼・前電通大 山崎 昶編

化 合 物 の 辞 典

14043-6 C3543　　B5判 1008頁 本体55000円

工業製品のみならず身のまわりの製品も含めて私達は無機、有機の化合物の世界の中で生活しているといってもよい。そのような状況下で化学を専門としていない人が化合物の知識を必要とするケースも増大している。また研究者でも研究領域が異なると化合物名は知っていてもその物性、用途、毒性等までは知らないという例も多い。本書はそれらの要望に応えるために、無機化合物、有機化合物、さらに有機試薬を含めて約8000化合物を最新データをもとに詳細に解説した総合辞典

東大 梅澤喜夫編

化 学 測 定 の 事 典
―確度・精度・感度―

14070-3 C3043　　A5判 352頁 本体9500円

化学測定の3要素といわれる"確度""精度""感度"の重要性を説明し、具体的な研究実験例にてその詳細を提示する。〔実験例内容〕細胞機能(石井由晴・柳田敏雄)／プローブ分子(小澤岳昌)／DNAシーケンサー(神原秀記・釜堀政男)／蛍光プローブ(松本和子)／タンパク質(若林健之)／イオン化と質量分析(山下雅道)／隕石(海老原充)／星間分子(山本智)／火山ガス化学組成(野津憲治)／オゾンホール(廣田道夫)／ヒ素試料(中井泉)／ラマン分光(浜口宏夫)／STM(梅澤喜夫・西野智昭)

日本分析化学会編

機 器 分 析 の 事 典

14069-X C3543　　A5判 360頁 本体12000円

今日の科学の発展に伴い測定機器や計測技術は高度化し、測定の対象も拡大、微細化している。こうした状況の中で、実験の目的や環境、試料に適した機器を選び利用するために測定機器に関する知識をもつことの重要性は非常に大きい。本書は理工学・医学・薬学・農学等の分野において実際の測定に用いる機器の構成、作動原理、得られる定性・定量情報、用途、応用例などを解説する。〔項目〕ICP-MS／イオンセンサー／走査電子顕微鏡／等速電気泳動装置／超臨界流体抽出装置／他

上記価格(税別)は2006年1月現在

別表 広く使用される略語と記号

用語	略語・記号	和訳，用例
about (circa)	ca.	およそ，ca. 1980, ca. 120 g
alternating current	AC, ac	交流電流，AC polarography
analysis	anal.	分析，Anal. Calcd for ...
and others (et alii)	et al.	およびその他の者
and so forth (et cetera)	etc.	など，その他
aqueous	aq	水の，M$^+$(aq), NaCl(aq)
atomic weight	at. wt	原子量，at. wt 162.50
body-centered cubic	bcc	体心立方
boiling point	bp	沸点，bp 256 ℃
calculated	calcd	計算された，D_{calcd}
chemically pure	CP	化学用（一般用試薬）
circular dichroism	CD	円偏光二色性
confer, compare	cf.	を参照，比較せよ
decomposition	dec	分解，mp 225℃ dec
deoxyribonucleic acid	DNA	デオキシリボ核酸
direct current	DC, dc	直流電流
electromotive force	emf	起電力
electron spin resonance	ESR	電子スピン共鳴
equation, equations	eq, eqs	式，eq 1, eqs 1–3
face-centered cubic	fcc	面心立方
for example (exempli gratia)	e.g.	例えば
formula weight	fw	式量，fw 104.2
freezing point	fp	凝固点
gas	g	気体，H_2O(g)
gas–liquid (partition) chromatography	GLC (GLPC)	ガス–液体（分配）クロマトグラフィー
guaranteed reagent	GR	保証付試薬
hexagonal close packed	hcp	六方最密充填
hyperfine splitting	hfs	超微細分裂
infrared	IR	赤外線の，赤外線
liquid	l	液体，H_2O(l)
literature	lit	文献，lit 165−166℃
mass spectrometry	MS	質量分光測定

用語	略語・記号	和訳，用例
mass spectrum	MS	質量スペクトル
melting point	mp	融点，mp 160 ℃
molality	m	重量モル濃度（mol/1000g 溶媒）
molarity	M	モル濃度（mol/1L 溶液）
molecular weight	mol wt	分子量，mol wt 395.4
nuclear magnetic resonance	NMR	核磁気共鳴
number, numbers	no., nos.	番号，no. 1，nos. 1–5
optical rotatory dispersion	ORD	旋光分散
parts per million	ppm	100万分の1，5 ppm
parts per billion	ppb	10億分の1，8 ppb
reference, references	ref, refs	参考文献，ref 1，refs 1–3
relative molecular mass	M_r	相対分子質量，M_r = 395.4
ribonucleic acid	RNA	リボ核酸
solid	s	固体，$H_2O(s)$
species (singular and plural)	sp., spp.	種名
that is (id est)	i.e.	すなわち
thin-layer chromatography	TLC	薄層クロマトグラフィー
ultraviolet	UV	紫外の
ultraviolet–visible	UV–vis	紫外可視の
United Kingdom	U.K.	連合王国（英国）
United States	U.S.	合衆国（米国）
versus	vs.	比較して，…対する
volume	vol	体積
volume per volume	v/v	体積比
weight	wt	重量
weight per volume	w/v	重量体積比
weight per weight	w/w	重量比